Photoelectric Sensors and Controls

MECHANICAL ENGINEERING

A Series of Textbooks and Reference Books

EDITORS

L. L. FAULKNER

Columbus Division
Battelle Memorial Institute

and

Department of Mechanical Engineering
The Ohio State University
Columbus, Ohio

S. B. MENKES

Department of Mechanical Engineering
The City College of the
City University of New York
New York, New York

1. Spring Designer's Handbook, *by Harold Carlson*
2. Computer-Aided Graphics and Design, *by Daniel L. Ryan*
3. Lubrication Fundamentals, *by J. George Wills*
4. Solar Engineering for Domestic Buildings, *by William A. Himmelman*
5. Applied Engineering Mechanics: Statics and Dynamics, *by G. Boothroyd and C. Poli*
6. Centrifugal Pump Clinic, *by Igor J. Karassik*
7. Computer-Aided Kinetics for Machine Design, *by Daniel L. Ryan*
8. Plastics Products Design Handbook, Part A: Materials and Components; Part B: Processes and Design for Processes, *edited by Edward Miller*
9. Turbomachinery: Basic Theory and Applications, *by Earl Logan, Jr.*
10. Vibrations of Shells and Plates, *by Werner Soedel*
11. Flat and Corrugated Diaphragm Design Handbook, *by Mario Di Giovanni*
12. Practical Stress Analysis in Engineering Design, *by Alexander Blake*
13. An Introduction to the Design and Behavior of Bolted Joints, *by John H. Bickford*
14. Optimal Engineering Design: Principles and Applications, *by James N. Siddall*
15. Spring Manufacturing Handbook, *by Harold Carlson*
16. Industrial Noise Control: Fundamentals and Applications, *edited by Lewis H. Bell*
17. Gears and Their Vibration: A Basic Approach to Understanding Gear Noise, *by J. Derek Smith*

Additional Volumes in Preparation

Mechanical Engineering Software

Spring Design with an IBM PC, *by Al Dietrich*

Mechanical Design Failure Analysis: With Failure Analysis System Software for the IBM PC, *by David G. Ullman*

Photoelectric Sensors and Controls

SELECTION AND APPLICATION

Scott M. Juds
Opcon
Everett, Washington

MARCEL DEKKER, INC.   New York and Basel

Library of Congress Cataloging-in-Publication Data

Juds, Scott
-- Photoelectric sensors and controls.

   (Mechanical engineering ; 63)
   Includes index.
   1. Photoelectronic devices. I. Title. II. Series:
Mechanical engineering (Marcel Dekker, Inc.) ; 63.
TK8304.J83 1988   621.3815'42   87-37985
ISBN 0-8247-7886-3

MARCEL DEKKER, INC.
270 Madison Avenue, New York, New York 10016

Current printing (last digit):
10 9 8 7 6 5 4 3 2 1

PRINTED IN THE UNITED STATES OF AMERICA

To my children:

Stephanie Louise
Eric B. Arthur

Preface

Light sensing as a means of industrial control has been available for many decades. The technology incorporated into photoelectric controls and sensors since the early 1970s has revolutionized the market with significant advances in the state of the art in electronics, optics, and packaging. These advances have enabled photoelectric controls and sensors to overcome most of the barriers to their practical application.

The scope of technologies encompassed by these products generally extends into one or more areas in which the user has little background or training. This book is designed to provide the background and reference information necessary for making informed selection decisions. It is designed to improve the success in application of these devices by maintenance personnel, electricians, designers, and engineers. Material in this text is targeted at system designers, who must make the initial selection, and at maintenance personnel, who make sure installed systems keep working.

The book covers the basic fundamentals of optics and also gives an in-depth practical analysis of the major sensor configurations. Electrical control interfaces, control logic functions and specifications are described. Environmental issues, often a "gotcha" and important for system reliability, are discussed.

Finally, application issues and specific examples are discussed to help the reader understand the possibilities, practical limitations, and pitfalls. I thoroughly recommend experimentation as you read to help solidify concepts. Confucius put it this way: "I hear and I forget, I see and I remember, I do and I understand."

The text was specifically designed to be useful to readers with minimal experience in enclosures, optics, electronics, or industrial controls. Many drawings and graphs are used to simplify explanations and impart a solid practical understanding.

After reading *Photoelectric Sensors and Controls* you will be prepared with the insight and knowledge to tackle even the toughest optical sensing and control problems.

I would like to extend special thanks and appreciation to physicist Rocky Kyle for his technical review and for the photoelectric optics expertise he was willing to share over the years; to electrical engineer and photoelectric expert Paul Mathews, who reviewed the manuscript while sailing his boat to exotic places in the South Pacific; to artist John Jolley for his accuracy and aesthetic interpretation of my stick figures; and to secretary Jeri Barnhart for her assistance with the intricacies of our written language, compilations, and correspondence.

Scott M. Juds

Contents

Contents

Contents

Photoelectric Sensors and Controls

1

Photoelectric Sensors
and Controls: Introduction

History, a distillation of rumour.

Thomas Carlyle
The French Revolution

In this chapter we provide a description of photoelectric controls
and sensors, what they are, principles of operation, basic terminology, a look backward, and a look forward. Much of the information here will already be familiar to many readers. However,
the foundation for material in other chapters will be laid here and
may be useful to a complete understanding of later chapters.

1.1 INTRODUCTION

The number and variety of light-operated or light-controlled devices and equipment produced is tremendous. Photoelectric controls and sensors are only a small part of this vast product spectrum. Dusk lamp controls, television brightness compensation
for ambient light, photo interrupters for paper and ribbon cart-
ridge empty signals on printers, cassette auto-reverse sensors,
automobile ignition electronic cams, television and stereo remote
controls, door annunciators in small shops, smoke detectors, CRT
light pens, the optical mouse for personal computers, shaft en-
coders, elevator door safety guards, invisible perimeter security
sensors, and garage and toll gate automobile detection are a few
of the many light-sensitive controls and sensors with which we
interact in everyday life. Many of these are very specific in
form to a particular market or need. As such, they will not be
discussed here directly, although the principles of operation are

1

likely to apply. Hidden within industry are countless other varieties of optically based sensors, controls, array scanners, bar code
scanners, measurement scanners, range finders, and machine vision systems. In the larger sense, all of the aforementioned devices could be considered photoelectric controls and sensors.
However, in this book, a more categorical definition will be used
with particular attention to industrial sensors and controls.

The term "sensor" will be used to refer to a device that provides a simple signal-level output that is determined by the light
level it receives. The term "control" will be used to refer to a
device that, in addition to sensing, provides a control output capable of switching power to actuators that alter the flow of materials or other process events. Both sensors and controls are
concerned with only a single sense point. The term "scanner"
will be reserved for devices that have built-in physical or electronically synthesized scanning motion in order to gather information from more than a single sense point. A scanner may be
a self-contained unit or an ensemble of sensors connected to a
scan controller. Single-point sensors and controls are often erroneously referred to as scanners despite their inability, physically or electronically, to move the sense point in space. Although fixed-beam sensors and controls are not scanners, they
can provide that function when the scanning motion is provided
external to the sensor by the object's motion through the sensor's beam. In this book, these definitions are used consistently
to avoid confusion.

1.2 PRINCIPLES OF OPERATION

An industrial photoelectric sensor is a device composed of a light
transmitter and light receiver. Light is directed toward the object by the transmitter. The receiver is pointed toward the same
object and detects the presence or absence of reflected light or-
iginating from the transmitter. Detection of the light generates
an output signal for use by an actuator, controller, or computer.
The output signal can be analog or digital and is often internally
modified with timing logic, scaling, or offset adjustments prior to
output.

1.2.1 Conventional Unmodulated Photoelectrics

The light beam in early photoelectrics was generated by a con-
stant-intensity (unmodulated) incandescent filament bulb much

like those used for automotive taillights. The light detector was
generally a cadmium sulfide cell that allowed current to flow and
energize a relay when lit, and blocked current flow, deenergizing
the relay when dark. A cadmium sulfide cell is a photoresistive
detector. Its ability to conduct or resist the flow of current is
controlled by the intensity of light incident on its surface. Early
photoelectrics were generally called "electric eyes" and even today
are frequently referred to as "eyes." Figure 1.1a shows an early
electric eye, and Fig. 1.1b shows schematically how it functioned.
In operation, this device projected light toward a reflector that
would return it back to the electric eye's cadmium sulfide cell.
If enough light fell on the cell, sufficient current would be conducted through the cell to activate the relay. It was generally
necessary to shield the detector from other light sources and to
use a powerful filament light source to achieve any sensitivity
without false triggering. An object breaking the beam would
then cause the darkened cadmium sulfide cell to switch off the
relay.

In the late 1960s, the light-emitting diode (LED) was developed
and provided an alternative light source to the incandescent light
bulb. LEDs are a common sight today as numerical displays in
clocks and stereos, illuminators in toy games, and indicator lamps
on security system controls and floppy disk drives. A solid-state
LED, sketched in Fig. 1.2, is constructed without the two most
serious life-limiting design problems found in incandescent bulbs.
Incandescent bulb tungsten filaments operate at about 4000°F.
This high operating temperature creates great thermal stress,
causing deterioration of the filament structure over time. The
high operating temperature limits the useful life of these bulbs
to about 2000 to 5000 hours. Further, the filament is a coil spring
structure stretched between two posts. This delicate structure is
stressed heavily by shock and vibration common to industrial production environments. The already short bulb life is significantly
reduced by shock and vibration. In contrast, the LED has a bond
wire that is encapsulated and is able to withstand extremely high
shock and vibration stress. The LED operates at nearly ambient
temperature and does not encounter great thermal stress. The
typical rated life of an LED is in excess of 100,000 hours, which
is more than 10 years of 24-hour days. This does not guarantee
that all LEDs will last 10 years and then die. It means that the
field failure rate will be on the order of 2 to 5% of that experi-
enced with incandescent bulbs. This is a feature of significant
value when the cost of maintenance and production downtime are
considered.

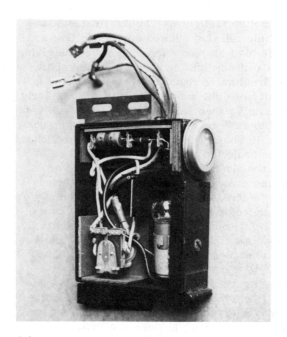

(a)

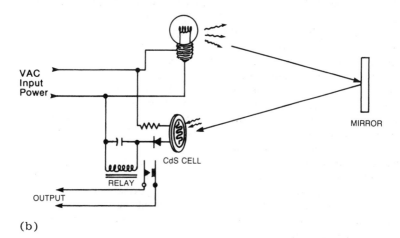

(b)

Figure 1.1 (a) Early electric eye with its cover removed; (b) schematic diagram of its function.

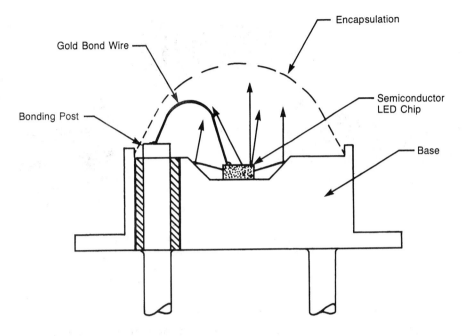

Figure 1.2 LED with a semiconductor chip mounted on the base. The connection of the LED to the bonding post with a tiny gold bond wire is encapsulated for mechanical protection.

Other advances made in conventional photoelectric product technology include the incorporation of lenses to form a collimated beam of light and the incorporation of the silicon phototransistor. The addition of collimating lenses improved performance in the presence of moderate indoor ambient lighting by limiting the sensor's field of view. The silicon phototransistor brought with it greater reliability, smaller size, and faster response time. Conventional unmodulated photoelectrics have simple direct responding circuits that are low cost and quite fast. Although their use is complicated by required attention to shielding from local ambient light sources, unmodulated photoelectrics still command a fair fraction of the photoelectric market today when lowest cost or extremely fast response times are required.

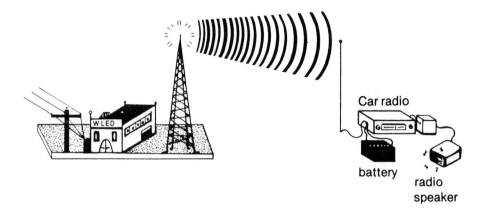

Figure 1.3 A radio transmitter and receiver are analogous to a
pulsed LED transmitter and receiver. (Courtesy of Opcon, Inc.)

1.2.2 Modern Pulse-Modulated Photoelectrics

The LED brought with it another important advantage. The LED
can be modulated (turned on and off) at very high rates. Modu-
lation allows the detector to have a method to distinguish between
the LED-produced pulsed light and ambient light. A pulse-modulated photoelectric sensor operates on a principle very similar to
that of the radio station and radio receiver shown in Fig. 1.3.
The receiver is tuned to accept the desired radio frequency and
reject possible interfering frequencies from other radio sources.
Similarly, there are many sources of light interference for photoelectric sensors, such as sunlight, incandescent lamps, fluorescent lamps, light interrupted by fans, arc weld flash, beacons,
and flames that must be rejected by the photoelectric receiver.
    Figure 1.4 shows how a photoelectric sensor is constructed to
use pulse-modulated LED light to obtain rejection of other ambient light sources. An oscillator circuit generates electrical pulses that cause an LED to generate pulsed light. The response
time of a cadmium sulfide cell is not fast enough to detect these
fast pulses. Instead, pulse-modulated photoelectrics use silicon
photodiodes or phototransistors that can respond fast enough to
resolve pulses as short as 10 µs (ten millionths of a second) in
duration. Receiver amplifier and detector circuits are designed
to respond optimally to the incoming pulsed light signals. Because most light sources produce no light modulated as high in

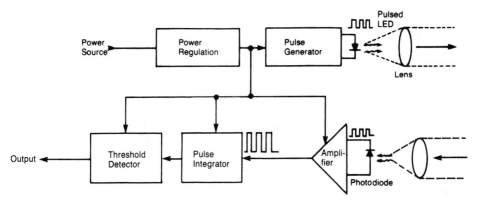

Figure 1.4 Pulsed LED photoelectric sensor block diagram.

frequency as that of the LED, it is not difficult to distinguish between them. However, a few light sources, such as strobe lights and arc weld flash, produce light pulses that may fool some of the more simply designed modulated photoelectric receiver circuits. In most applications the gain in performance with pulse-modulated LED technology is tremendous. Although sunlight falling on a box to be detected is about 100,000 times stronger than the illumination from a pulsed LED, most of today's sensors are capable of detecting the box in the presence of such strong interfering illumination. The principal advantage of pulse-modulated photoelectric sensors is their ability to sense only the light emitted by the pulsed LED while rejecting other sources of light that would normally interfere with the operation of a conventional sensor. The result is extremely high optical performance and repeatability in the face of changing lighting conditions.

LEDs come in a number of colors. Recently, blue has been added to the already available LED colors of green, yellow, orange, red, and infrared. The physics of solid-state light generation and detection favors the use of the longer infrared wavelengths for photoelectrics. The available optical performance of infrared sensors is 10 to 100 times better than that of their visible counterparts. However, when detecting color-contrasted materials, there are significant advantages to visible color sensors that often outweigh the optical strength disadvantage.

1.3 PRINCIPAL OPTICAL MODES OF OPERATION

Three basic methods of sensing are used in photoelectrics: thru-beam, reflex, and proximity, each of which operates on the same electronic principles. The primary difference is in how the optics are arranged. They are each designed somewhat differently in order to optimize the interaction of the light beam and target for optimum performance in the particular sensing mode. Within each of these three basic sensing modes, there are minor variations that give performance advantages for specific applications. These are covered in detail in Chapter 3; here, the three modes are introduced briefly.

1.3.1 Thru-Beam Sensing

Thru-beam sensing is the oldest and most familiar type to most people. It appears often in movies and television programs as invisible light beams forming a secured perimeter for an alarm system protecting a valuable art object, jewel, or top-secret building. These sensors are sometimes referred to by such names as "break-beam," "thru-scan," or "opposed mode." As shown diagrammatically in Fig. 1.5, a thru-beam sensor consists of two parts: a pulsed light source and a pulsed light detector. The light source is commonly called the source, emitter, sender, or transmitter, and the light detector is commonly called the detector or receiver. In normal operation, the beam is complete or uninterrupted between the source and detector. Source light pulses are detected until an object interrupts or breaks the light beam. The loss of signal results in a change in the sensor output. Because the detector is aimed directly at the source rather than the source's reflection, thru-beam sensors have the longest-range capability of the three modes. As such, they are commonly

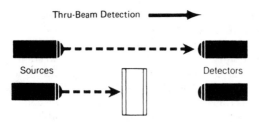

Figure 1.5 Thru-beam sensor configuration.

used for perimeter security, elevator door guards, and across shipping docks.

Features of Thru-Beam Detection
    Longest optical range
    Highest possible signal strength (excess gain)
    Greatest light/dark contrast ratio
    Little effect by surface color and reflectivity
    Best trip-point repeatability
Limitations of Thru-Beam Detection
    Two components to wire across detection zone
    Possible difficult alignment

1.3.2 Reflex Sensing

Reflex sensing is the most popular method used. "Reflex" means to bend or turn back. Reflex sensors use retroreflective targets that causes the incident light to be folded back, to return along the path from which it came. These sensors are generally called either retroreflective or reflex sensors. The reflex mode has an advantage over the thru-beam mode since it does not require that wires be run to a second location. A reflex sensor is shown diagrammatically in Fig. 1.6. In use, these sensors operate much the same as a thru-beam sensor. An object of low reflectivity interrupts the beam to the retroreflector. Retroreflectors were

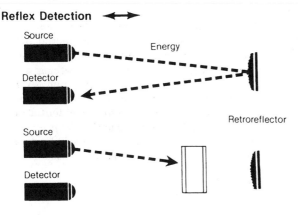

Figure 1.6 Reflex sensor configuration.

developed to make highway signs highly visible by bending light
back to the cars from which it came rather than scattering it in
all directions. Compared to an ordinary white paper surface, inexpensive plastic retroreflectors return about 5000 times as much
light back to the vicinity of its origin. This provides for high
contrast and is why interruption of a reflex beam with cardboard
boxes, wood, vehicles, and other objects is a reliable sensing
method. Reflex sensing is a powerful sensing method that does
not require electrical wire to be run to both sides of the sensing
area. The combination of high sensing power and ease of installation has made reflex sensing the most popular choice of all sensing
modes.

Features of Reflex Detection
   Long optical range
   High light/dark contrast ratio
   Little effect by surface color and reflectivity
   Easy installation and alignment
Limitations of Reflex Detection
   Can be false triggered by mirrorlike objects unless polarized
     optics are used

1.3.3 Proximity Sensing

Proximity sensing is accomplished by detecting light returned directly from an object's surface. Proximity sensing is sometimes
referred to as diffuse proximity or diffuse scan. Figure 1.7 shows
diagrammatically that diffusely reflected light is scattered back toward the sensor and detected. This method is attractive because
there is no need to locate a receiving sensor or retroreflector at
the other end of the light beam. Because light is reflected hap-
hazardly by the sensed object, proximity sensors have the short-
est detection range. Long-range proximity photoelectrics are
limited to about 10 ft of sensing range with a white paper target.
Other varieties of these sensors have been designed to provide
defined or limited range sensing in order to avoid accidental detection of background objects. Still other varieties have been developed with wide fan or pinpoint beam shapes. Carefully chosen
and applied, they can solve a variety of special problems.

Features of Proximity Detection
   Wires on only one side of the sensing zone
   Simple alignment
   Can detect differences in surface color and reflectivity

Proximity Detection

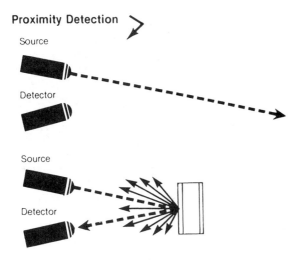

Figure 1.7 Proximity sensor configuration.

Limitations of Proximity Detection
    Limited sensing range
    Sensing range dependent on surface reflectivity of target

1.4 PHOTOELECTRIC TRENDS

As in all modern equipment, new technology is incorporated into
new products by manufacturers as it becomes available and cost-
effective. As markets mature, industrial standards and de facto
standards evolve out of user preference as determined by a product's ability to be the most cost-effective solution to a particular
class of sensing problems. Such standards include environmental
operating conditions, packaging and mounting styles, function
commonality, terminology, and product technology. As industrial
production methods and technology change, new demands are
placed on what used to be an acceptable product solution. In
Chapter 6 we address environmental testing standards and packaging standards that have evolved and pertain to products manufactured and used today.

1.4.1 Circuit Technology

The most significant catalyst for product technology improvement
has been the rapid advance in electronic technology and its continued cost reduction in the face of inflation. In constant dollars,
photoelectric controls cost about 25% of what they did 15 years
ago for the same performance. As technology improves, we tend
to desire and expect the improvements that come with it. The
cost/performance ratio of photoelectric sensors today is truly a
bargain and will be even more so in the future. Figure 1.8 demonstrates how electronics assembly has become both miniaturized
and more complex. The construction in Fig. 1.8a is called point-
to-point wiring. It is clearly a manual and space-consuming
method. This method was used in production electronics into the
1960s. Printed circuit board technology, shown in Fig. 1.8b, allows for automatic placement and machine soldering of components.
The components are inserted into a printed circuit board that has
a photochemically produced conductor interconnection pattern on
its surface. This technology is still the most commonly used in
the United States. Printed circuit boards are easily handled, are
easily repaired, and make nice functional circuit modules. Printed
circuit board technology is being replaced by surface-mount technology, shown in Fig. 1.8c, and already dominates production in
Japan and some European countries. Surface-mountable components have no lead wires. The reduction in size of the electronic
components reduces the size of circuit board material and packaging. The circuit board for surface-mount technology is produced either by the same methods as those used for a standard
printed circuit board, or via silk screen, printing onto its surface the oven-cured conductors, insulators, and resistor components. Surface-mount assembly for many high-volume electronic
products is completely automated. The greatest reduction in size
with simultaneous increase in performance has come from integration on a single silicon chip of most of the complex pulse modula-
tion and demodulation circuitry. Literally hundreds of circuit
components can be put onto a silicon chip no larger than 0.080
in. by 0.080 in., as shown in Fig. 1.8d. Many photoelectric manufacturers have already integrated one or more of their complex
circuits. These custom circuits may be mounted and wire bonded
directly to a surface-mount circuit board, they may be placed into a plastic package for surface-mount soldering, or they may be
placed into a standard leaded integrated circuit (IC) package for
use on printed circuit boards. Because of the distinctly different technologies, voltages, or power dissipations required in

components such as the LED, filter cpaacitors, and output devices, these components will not likely be integrated onto the same silicon chip in the near future. However, the trend toward silicon integration and surface-mount technology will, for some time, continue to drive down the package size while retaining or improving sensing performance.

1.4.2 Interference Testing

The new industrial environment has new demands. Electromagnetic interference (EMI) and radio-frequency interference (RFI) are hot topics today. Radio-controlled equipment or walkie-talkie communication gear is commonly used by plant maintenance personnel today. Most new electronic control equipment has high-frequency switching power supply converter circuits or other electrical circuits that generate interference noise. Some equipment even uses the 115 VAC power lines for communication by putting a high frequency carrier signal on top of the 115 VAC. These EMI and RFI sources are in addition to the showering arc noise generated by the opening of power contacts driving inductive loads such as solenoids and motors. The test methods described in Chapter 6 provide a fair confidence level of EMI and RFI immunity for sensors passing these tests. EMI and RFI noise has been described as a "creepy-crawly-black-magic-sort-of-stuff," because it is poorly understood by most people, is hard to measure, and is difficult to stop or fix after a problem is found. As manufacturers have become aware of the seriousness of this problem, some have built products clearly designed to pass stringent noise immunity tests. The need for circuit immunity to such electrical noise sources will continue to be important. More attention is likely to be paid to product specification and design in this area as better standards appear.

1.4.3 Interfacing

A significant trend in industry is the widespread use of the programmable logic controller (PLC) and industrially hardened minicomputers and microcomputers, which are playing a larger role in industrial automation. This has created an increased market need for simple dumb sensors which have no need for power output circuits, built-in timing logic, or light/dark output selection. All of these functions are taken care of by the PLC or industrial computer. These extra functions are required when the photoelectric sensor must directly control and operate a device such

(a)

(b)

Figure 1.8 The evolution of shrinking photoelectrics has gone
from (a) point-to-point hand wiring to (b) printed circuit boards
to (c) surface-mount technology, to (d) custom integrated circuits designed for photoelectric sensors. (Courtesy of Opcon,
Inc.)

(c)

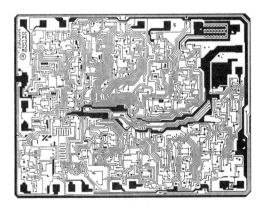

(d)

as a reject mechanism. There will always be a requirement for direct local control of devices by photoelectrics. However, simple DC-powered sensors will continue to gain substantially in market share as factory automation increases. Elimination of the provision for timing logic and high power outputs will result in further reduction in product size and cost.

1.4.4 Optics

Most of the present improvements in photoelectric optical performance are being made in the areas of miniaturization and alignment aids for the user. Alignment or signal strength indication is a feature designed to help facilitate the optimum installation and optical alignment of a photoelectric sensor. At this time, most available photoelectrics have only a single LED indicator that is either on or off, depending on the presence or absence of returned pulsed light. It is important to have good signal strength to avoid intermittent failures when a little dust builds up on the lens or the sensor is bumped a little further out of alignment. Products are offered by a few manufacturers that have begun to address the ease-of-use issue by offering built-in signal strength indicators. There are four methods currently in use: (1) The LED indicator brightness can be increased with signal strength, (2) the LED indicator can be caused to flash at a rate proportional to signal strength, (3) a bicolor LED can change its color as the signal strength increases, or (4) a multiple-bar LED level indicator can change the lighted LED position with signal strength. Signal strength indication will become easier to implement as these circuits are integrated onto silicon chips, and as a result they can be expected to appear in a larger proportion of future products.

The visible-beam LED will continue to become more prevalent as the light-conversion efficiency of the LED semiconductor process increases and demodulation techniques improve through circuit integration. Visible-beam devices are often easier to align with reflex and thru-beam sensors because the beam can be seen when it hits a retroreflector or can be observed directly at the far end of the beam. Visible LEDs perform much better with low-cost plastic fiber cables than do infrared LEDs, and thus will be the prevalent technology with those fibers.

1.4.5 Packaging

The trend in product packaging is toward smaller plastic housings which have been internally potted to provide additional mechanical strength and higher shock and vibration tolerances. Engineering plastics have been tremendously improved since the brittle Bakelite of the 1940s. The next stride in packaging technology,

which is just emerging, is molding of the case and filler around
the electronics in a single step. Although this construction method
virtually prevents repair of a defective sensor, the cost of these
products will be low enough that replacement will be cheaper than
repair.

2

Optical Fundamentals

*Facts, or what man believes to be facts, are delightful...
Get your facts first, and then distort them as much as
you please.*

Mark Twain
(Kipling, *From Sea to Sea*, Letter 37)

This chapter is intended to provide the background necessary
for an understanding of how light interacts with lenses, diffuse
reflecting objects, mirrors, retroreflectors, colored objects,
transparent and semitransparent materials, and photoelements.
A basic understanding of these components and materials will
lead to a better understanding of how different photoelectric
optical systems can be expected to react with the targets and
contaminants in their operating environment. So let's become
armchair physicists for the next few pages. Application problem solving through "intuitive physics" is your goal and will result in a lot of personal satisfaction when achieved.

2.1 PRODUCING THE BEAM

Many photoelectric models are manufactured in the same physical package and use the same electronic circuitry, yet one model
is a reflex sensor and the other is a proximity sensor. The difference rests in the specific choice of photoelements and their
location with respect to the lens. This small change creates different beam shapes and orientations that are responsible for significantly different sensing properties.

2.1.1 Photoelements

Photoelements are the electronic sources and detectors of light.
In modern photoelectric sensors, the light source is a pulsed
light-emitting diode (LED). An LED is a tiny chip of semiconductor crystal made of materials such as gallium arsenide, gallium aluminum arsenide, or gallium phosphide. This material is
fabricated to form a junction diode with the property of electro-
luminescence. When a current is caused to flow through the
junction diode, it produces light. Different material combinations cause emission of light of different colors. LEDs are packaged in a variety of ways, as indicated in Figs. 1.2 and 2.1.
Photodetectors are generally made of a silicon junction diode
that generates current when light strikes it. Solar cells used
for power generation from sunlight are just large versions of
these photodetectors.

A typical LED measures about 0.015 in. on a side. Upon very
close inspection of this device, a small wire can be seen connecting the terminal bonding post to the bonding pad in the center
of the LED, and a reflecting dish can be seen surrounding the

(a)   (b)

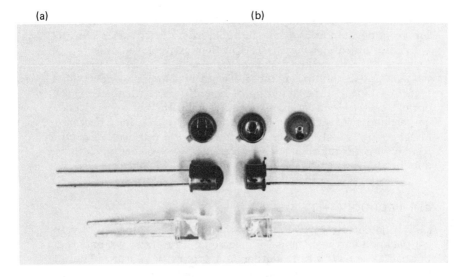

Figure 2.1 Typical 0.2-in.-diameter photoelements used in
photoelectric sensors (a) with, and (b) without immersion
lenses.

LED chip. When the LED is producing light, these objects will alter the light pattern projected by the sensor through the main lens by blocking or reflecting light. The wire will block some of the light over the LED chip, and the reflecting dish will add a ring of light around the LED chip. A typical photodetector chip is about 0.040 in. on a side. It generally will be wire bonded in a corner and will not have a reflecting dish surrounding it. Some of these features can be seen in Fig. 2.1b.

2.1.2 Lenses

Photoelements sometimes have a small integral lens built into the protective window. This is called an immersion lens because the lens often fully immerses the semiconductor photoelement. It is used in some photoelectric sensors to magnify the apparent size of the photoelement. In some sensor configurations, the immersion lens will improve the ease of alignment or simply direct more light energy toward the target to improve overall sensor sensitivity. Figure 2.1a illustrates how the immersion lens magnifies the apparent size of the photoelements. The immersion lens of the photodetector is completely blackened by the magnified image of the photodiode chip.

The large lenses common on photoelectric sensors direct light energy toward a target and collect any returned energy. These lenses are referred to as objective lenses. They provide the same function for the sensor as does the lens on a slide projector or a telescope. Lenses are used to focus or image light from one object to another. In photoelectric sensors, light is emitted by a source LED toward a lens that bends its rays and focuses them toward some distant target on the other side of the lens. Light striking the target may be partially reflected or otherwise transmitted back toward the sensor, where the detector's light collecting lens in turn will bend the incoming rays and focus them onto the photodetector. The simple analogy to the lens systems of a slide projector and telescope clarifies these principles. The light pattern projected by the slide projector will only illuminate objects in the light path. Dark areas in the light pattern cannot illuminate the target. Similarly, there is only a limited field of view available when looking through the telescope. Only objects in this field of view having light cast on them may be detected. Different lens-photoelement arrangements are used to obtain different optical performance characteristics.

There are two principal ways in which photoelements are focused in photoelectrics: (1) focused at infinity, and (2) fixed focused at close range. Most often they are focused at infinity.

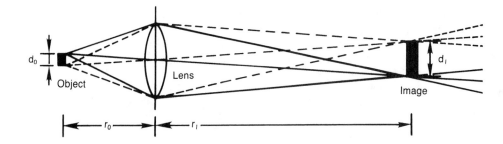

Figure 2.2 Fundamental relationships between the lens, object, and image of the object.

Figure 2.2 and Eqs. 2.1 and 2.2 show the fundamental lens relationships and equations for calculating focal distances and magnification:

$$\frac{1}{f} = \frac{1}{r_o} + \frac{1}{r_i} \qquad\qquad (2.1)$$

where f is the lens focal length; and

$$d_i = \frac{r_i}{r_o} d_o \qquad\qquad (2.2)$$

where r_i/r_o is the magnification factor.

Let's take a look at a numerical example just to get a feeling for how this works. It is not important that you learn to use these equations fluently. However, a real example now may help your intuition later. Using Eq. 2.1 and substituting infinity for the distance to the image (r_i), the object distance (r_o) and the lens focal length became one and the same. Hence this is the distance from lens to photoelement in most photoelectrics. Even if we wanted best focus at 24 in., using the focal length as the mounting distance is still an excellent approximation. If we plug in 24 in. and assume that the lens focal length is 0.75 in., optimal lens-to-photoelement mounting distance would be 0.774 in., hardly a difference worth thinking about. In photoelectric sensing, the question is "How much of the energy falls on the detector?" not "Are the edges of the image a little fuzzy?" Other optical configurations offer what is referred to as fixed-focus or convergent beam optics. The position of the lens and photoelements cause the light to be focused sharply in a small spot at a

Figure 2.3 The photoelectric control on the left is focused at
infinity. The one on the right is fixed focused at 4 in. and requires the extended lens mounting.

very short range. If the same lens is to focus the photoelement
image 4.0 in. in front of the lens, Eq. 2.1 requires that the lens
mounting position be extended to 0.923 in. from the photoelements. This extended lens is characteristic of most fixed-focus
proximity photoelectrics. Figure 2.3 shows both a photoelectric
focused at infinity and one focused at 4.0 in. with an extended
lens. Using Eq. 2.2 we can calculate the size of the spot the
LED will illuminate at 4 in. Assuming that the LED chip is
mounted in a reflecting dimple of diameter 0.040 in., we then
calculate the LED image to be a spot $(4.0)(0.040)/(0.932) =$
0.173 in. in diameter.

2.1.3 Beam Patterns

The beam pattern of LED light emerging from a lens and the field
of view of the detector may be thought of as one-and-the-same
mathematical relationship. The only difference is in the direction the light is traveling through the lens. Referring to Fig.
2.2, we can see that there is a very precise way to calculate the
size of the focused image projected by the LED through the lens.
For photoelectric sensors focused at a short distance, Fig. 2.2
implies that the light beam is as large as the entire lens at the
lens's surface and narrows to the smallest size when the image
is in focus. At distances larger than this, the light beam diverges. Figure 2.4 shows how this may be redrawn for the case
of focus at infinity. In the near field, close to the lens, the
beam diameter is the same as the lens diameter. In the far field,
usually at least 1 ft out, the beam diameter is controlled by the
size of the magnified image. The distance at which transition
from near field to far field occurs will depend on the size of the
photoelement, the diameter of the lens, and the focal length of
the lens. This distance may be calculated by Eq. 2.3 and is
shown diagrammatically in Fig. 2.4:

$$\text{Transition range } r_t = \frac{d_l}{d_p} f \tag{2.3}$$

where

   d_l = diameter of the lens
   d_p = diameter of photoelement
   f = lens focal length

For most photoelectrics the transition range can be estimated
quite easily by inspection. The lens diameter can be measured
directly. The focal length (f) is usually about 1.5 times the diameter of the lens. Retroreflective sensors usually use an LED
about 0.015 in. square and a photodiode about 0.040 in. square.
Photoelements in a diffuse proximity sensor will typically appear
0.15 in. in diameter. These rules of thumb are handy when beam
size must be known due to space restrictions in the sensing region. Hopefully, these data are available directly from the manufacturer's data sheet and need not be calculated. Alternatively,
contacting an application engineer at the factory may be the
easiest and surest route.
      Another way to think of the light beam is in terms of the
angular radiation pattern. In the far field, the beam diameter

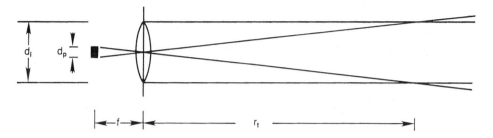

Figure 2.4 The range at which the diameter of a beam focused at infinity transitions from being controlled by the lens diameter to being controlled by the expanding image size is r_t.

expands as the distance from the lens increases. The ratio of the beam diameter to the distance from the lens will be constant and is usually expressed in degrees of beam divergence. The ratio of focal length to lens diameter in most photoelectrics is about 1.5:1 because of physical limitations of lens quality at smaller ratios, and decreased light coupling with larger ratios. If we hold this relationship constant and plug numbers into Eq. 2.2, we find that as the lens diameter and focal length increase, the LED image at a given range in the far field becomes smaller. A smaller image made with a larger lens is composed of the same amount of light as a larger image made with a smaller lens. The general result is that a larger lens will concentrate the same amount of light into a smaller angular radiation pattern. Figure 2.5 shows the relative radiation pattern for several lens diameters used with an LED mounted in a reflecting dimple. The reflecting dimple adds the side lobe bumps.

Figure 2.6 shows how the cross section of an LED light beam changes as the distance from the lens increases. The first photograph (a) is taken near the lens. It shows that the shape of the lens used was a square about 0.3 in. on a side. There are a few lines parallel to some of the sides that are artifacts of the lens edge construction. The second picture (b) was taken at the transition range. At the transition range, the beam is still about the same size but is a mix of the lens shape and the image of the LED that is beginning to form. The third picture (c) was taken in the far field, where the image of the LED is actually formed. In the far field all the structure related to the LED and its mounting become apparent.

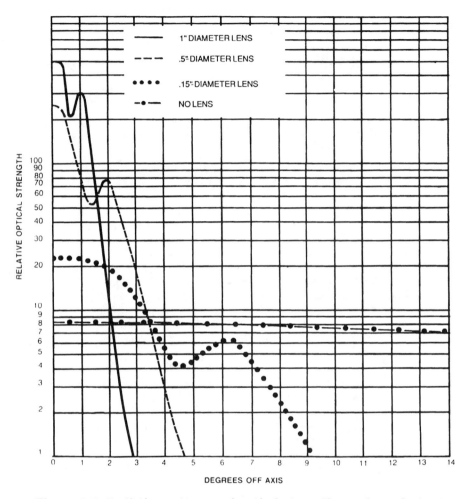

Figure 2.5 Radiation pattern and optical strength are dependent on lens diameter. The side lobe bump is present only when an LED is mounted in a reflecting dimple.

There are two factors that determine a proximity or reflex photoelectric's sensing region: The spatial intensity structure of the LED beam, and the region of source-detector beam overlap.

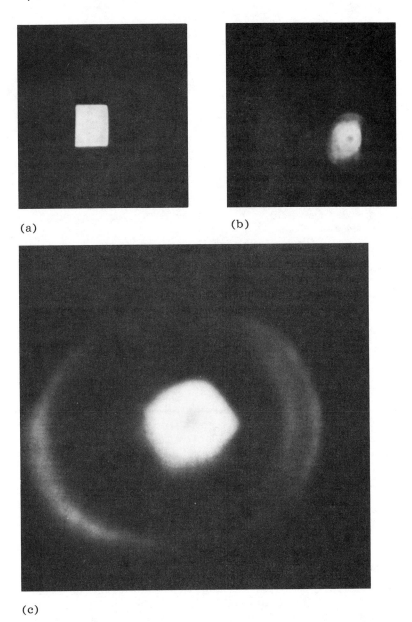

(a)

(b)

(c)

Figure 2.6 Projected image of the LED: (a) near the lens; (b) at the transition range r_t; (c) in the far field.

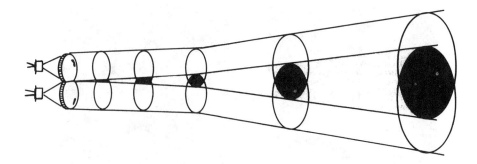

Figure 2.7 The active sensing region is only where the LED
beam and detector field of view overlap. This region is called
the "effective beam."

Only in places where the beams of both the source and the detector overlap will signal be returned when a target is in position. Similarly, flashlights and automobile headlights do not
produce a beam of uniform intensity. Objects can be seen only
if the light beam hits the object and the viewer's visual field of
view includes the object. Most people have trouble seeing things
in the dark or where they are not looking; so do photoelectric
sensors. In Fig. 2.7 the darkened region represents the overlap region for the sensor. It is often referred to as the "effective beam." Most photoelectric controls use a merged double
lens, as depicted in Fig. 2.3. Figure 2.7 shows that near a
merged double lens, the only place a target could be detected
is in a vertical stripe running the length of the line that joins
the two side-by-side lenses. As the distance from the lens increases, so does the area of the effective beam. In Chapter 3
we see how an effective beam can be tailored to obtain a variety
of useful sensing characteristics.

2.2 REFLECTION

Reflection is the return of radiation from a surface at the same
wavelength as the incident radiation. Specifically, this excludes
reradiation via fluorescent or thermal conversion. Reflection may
be diffuse, from a rough surface; specular, from a very smooth
surface; or a combination of both. Retroreflection, a special

class of specular reflection is of particular interest to photoelectric sensing.

2.2.1 Diffuse Reflection

A diffuse reflector is one with an optically rough surface that scatters light in all directions. Consider that a light ray, being very tiny, interacts with only a microscopic part of an object. Rough-textured surfaces present a variety of surface orientations to incident rays. The reflections from the surfaces will then also be in a variety of directions. Cotton is an excellent example of a diffuse reflector. Although light is reflected in all directions from a diffuse surface, light intensity is not constant in all directions. The intensity is greatest directly above the illuminated spot and diminishes as the viewing angle to any side increases. The diffuse reflection function, depicted in Fig. 2.8, is referred to as Lambertian reflection, after Heinrich Johann Lambert, who characterized the function as being proportional to the cosine of the angle between the surface normal (a line perpendicular to the surface) and the path of the radiant emission:

$$\text{Lambertian intensity } I = I_o \cos \theta \qquad\qquad (2.4)$$

This function holds true even if the incident light is not perpendicular to the surface. Microscopic examination of materials such

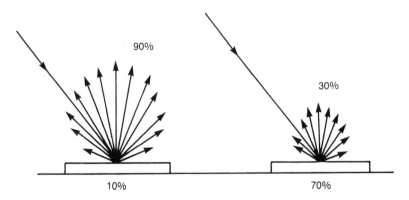

Figure 2.8 Diffuse reflection from partially absorbing Lambertian surfaces. (Courtesy of Opcon, Inc.)

as cloth, paper, and salt will reveal that these white things are actually made up of transparent fibers or crystals. These fibers and crystals act as little lenses and prisms that bend and rebend the incident light until virtually all of it is redirected back out of the surface. Because of the random orientation and number of fibers or crystals the light goes through, it has equal probability of exiting the surface at any angle. Although this might seem a contradiction with Lambertian reflection, it really is not. As the view angle from the normal grows, the apparent area of the radiating surface element diminishes as the cosine of the angle, thus effectively reducing the light contribution from the individual surface element. For example, as a coin is tipped from a view of its face to a view of its edge, the observed area of the face gets smaller and smaller until it is not seen at all.

2.2.2 Specular Reflection

The surface of most objects have at least a small amount of shine. In comparison to a rough diffuse reflecting surface, polished surfaces are smooth even on a microscopic scale. They reflect light quite consistently overall in the same direction, producing the visual effects of mirror reflections, glare, sheen, and luster. The general technical term is "specular reflection." Most of us are particularly familiar with the nearly perfect specular reflection of a mirror. Figure 2.9 illustrates that the angle of specular light reflection is the same as the angle of the incident light. By reflecting all rays at angles equal to their angle of incidence, images are preserved through the process of specular reflection. Most specular reflectors, however, are not nearly of the quality of a common silvered mirror. The two most significant imperfections of specular reflectors are partial reflections and curved surfaces. Partial specular reflections are most common among non-metallic surfaces. For example, although most light is transmitted by glass, the front and back surfaces of glass each specularly reflect about 4% of the incident light. Some of us have on occasion used windows, china plates, and similar items as mirrors even though the coefficient of specular reflection is quite low. Curved specular surfaces are a familiar sight on glossy paper stock, wet roads, drinking glasses, polished leather, lakes with ripples, antiglare glass, and many other objects. The curved surfaces cause the reflected image to be distorted. A single large curve will compress or expand the image size. Many small curves or bumps will slightly diffuse or blur the image by producing many scattered little images or image fragments, one

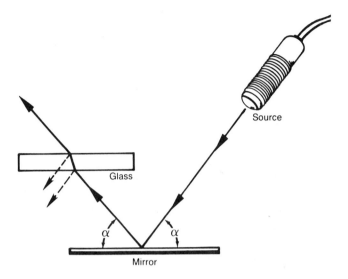

Figure 2.9 Specular reflection from a mirror and from glass.
The angle of reflection is equal to the angle of incidence.

from each bump and dip where the angle of reflection is the same
as the angle of incidence. On the microscopic scale, the angle
of reflection still equals the angle of incidence. However, because the microscopic surface is not quite flat relative to the
macro surface, the macro effect is a broadened specular peak
where the angle of reflection is nearly the same as the angle of
incidence.

2.2.3 Mixed Diffuse and Specular Reflection

Few objects are strictly diffuse or specular reflectors. Many objects, such as the glossy paper used in magazines, have both
specular and diffuse components in their reflection, as illustrated
in Fig. 2.10. The sensing range of a diffuse proximity optical
sensor is significantly increased in the presence of specular reflection. Depending on the application, this could be beneficial
if repeatable, or detrimental if not. For example, if a diffuse
proximity sensor normally detected a white piece of paper out to
30 in., it may be possible for the same sensor to detect the specular reflection from plate glass over 10 ft away. Because the
human eye is an auto-iris contrast-driven sensor, we make poor

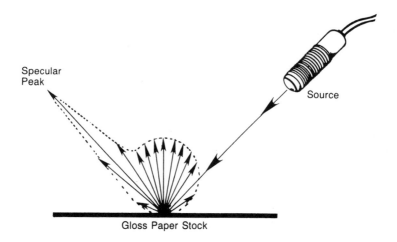

Figure 2.10 Mixed reflection from glossy paper stock contains both diffuse Lambertian reflection and specular reflection.

judges of the light intensity in reflections and should use experimentation rather than speculation when selecting a reliable optical sensing method. Figure 2.11 shows that common objects have both specular and diffuse components in their reflections. It should be noted that the materials on the chart may measure differently depending on the method of measurement, particularly when broadened specular peaks are involved.

2.2.4 Retroreflection

Retroreflection is the reflection of light by a reflector back in the direction from which it came. Reflectors specifically designed for this function are called retroreflectors and offer significant advantages over photoelectric sensors in terms of the light signal strength returned and ease of alignment. Photoelectric sensors designed to take advantage of this reflector type are called retroreflective or reflex sensors. "Reflex" means to bend, turn, or fold back.

One type of retroreflector is the corner cube reflector. Its operation can easily be understood by first observing how a two-dimensional corner reflector works. Following the incoming ray in Fig. 2.12, it undergoes two specular reflections. Recalling

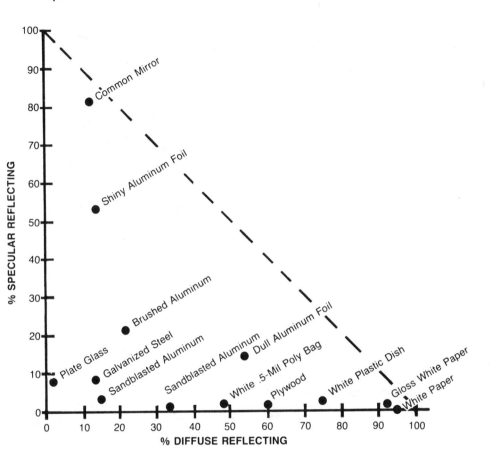

Figure 2.11 Specular and diffuse reflectance of various common objects.

that in specular reflection the angle of reflection is equal to the angle of incidence, we can calculate the angle of departure from the corner reflector. Adding up the two interior angles of the reflections, or all of the exterior angles of the reflections, results in a sum of 180°. A change in direction of 180° is by definition retroreflection. Similarly, the three-dimensional corner cube reflector of Fig. 2.13 can be analyzed with a little more trouble and shown to exhibit the same characteristics in three-

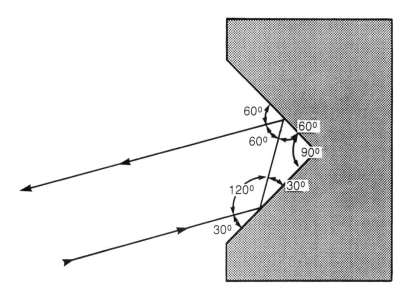

Figure 2.12 A two-dimensional corner reflector specularly reflects light twice, bending it a total of 180° to send it back in the direction from which it came.

dimensional space. Two types of corner cube reflectors are made. The original type is injection-molded acrylic plastic, is about 1/4 in. thick, and has about 150 corner cube prisms per square inch. Typical applications include as markers along the highway and at ends of driveways and as part of automobile taillight lenses. The second type, microprism reflectors, consist of tiny corner cube prisms bonded to or embossed in thin supporting films that may be as thin as 0.005 in. The reflector shown in Fig. 2.13, for example, consists of a prism layer less than 0.003 in. thick and has about 47,000 prisms per square inch. The quality of the retroreflection is related to the accuracy of the angles between the cube facets, the optically molded flatness of the facets, and the prevention of dust collection and water droplet formation on either side of the reflector (front or rear faceted sides). A typical plastic retroreflector will return approximately 5000 times the amount of light of a white diffuse reflector of the same size. This is why it is not difficult to make photoelectric sensors see a retroreflector and be blind to an ordinary cardboard box.

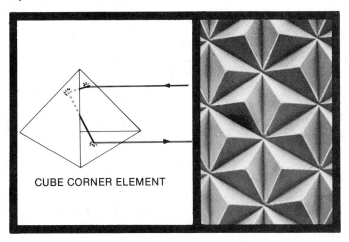

CUBE CORNER ELEMENT

Figure 2.13 (Left) A three-dimensional corner cube reflector uses the same principle as the two-dimensional corner reflector to return light in the direction of its source. (Right) A 100X magnification of a molded plastic corner cube array illustrates the precise placement of corner cube prisms in a typical retroreflector. (Courtesy of Reflexite Corporation.)

The other common retroreflector material is composed of microminiature glass beads, as illustrated in Fig. 2.14. The glass properties are chosen such that parallel rays of light entering the surface of the beads are focussed by the lens properties of its surface to a single spot on the other side of the bead. The bead is embedded in a diffuse reflecting binder. The Lambertian reflection in all directions travels back into the bead. Some of the light is scattered and rescattered into the sidewalls of the bead. However, the light that was reflected back toward the surface of the bead is then recollimated by the bead's lensed surface and travels back in the direction of the original illuminating source. Beaded retroreflectors are quite common in painted lines on highways, traffic signs, highly reflective tapes, and running shoes. Beaded retroreflection is typically 200 to 1000 times greater than that of a similar white diffuse reflector. Beaded retroreflectors are less effective than corner cube retroreflectors, for a number of reasons. A fair portion of the light is scattered by the binder between the beads and by a second or third encounter with the

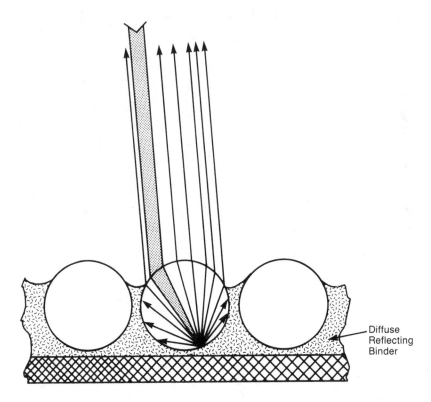

Figure 2.14 A beaded retroreflector focuses incoming energy on-to the reflective binder. The lens action of the bead redirects the scattered light reflection back toward its original source.

binder as it travels through the bead. This light will be scat-tered as regular Lambertian reflection. The rest of the light, which goes in and out of the beads as desired, may not be re-turned in exactly the same direction because of lens spherical aberration, refractive index variation, or aspherical beads; all of these effects cause a broadening of the angular distribution of the reflected light.

Figure 2.15 shows how the returned light energy of retrore-flectors quickly diminishes as the angle between illumination and observation increases. Both the plastic corner cube reflector array and the glass-beaded tape reflect about the same amount of light. However, glass-beaded tapes begin by losing over half

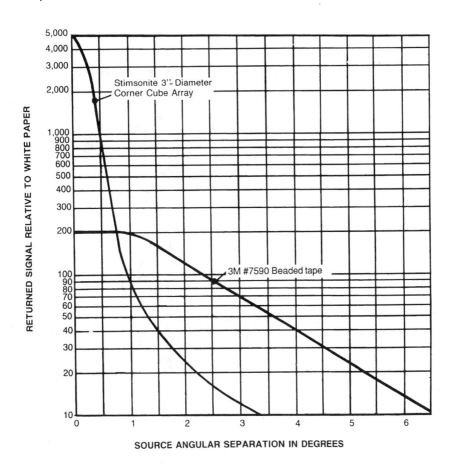

Figure 2.15 Most energy is returned by a retroreflector within a very small angle of the incident energy.

of their light to randomly directed diffuse reflections. Then, with regard to the retroreflected light, plastic corner cube reflectors appear brighter because they concentrate the same amount of light into a smaller angular spread. It should be pointed out that a perfect retroreflector would be a useless item for photoelectric sensors and the highway department because all of the incident light would be reflected directly back onto its source only. With both photoelectric sensors and automobiles, the observer or detector is adjacent to the light source. A road sign

at 100 yards has about a 0.2° observation angle between the
automobile headlight illumination and the driver's eyes. A retroreflector 3 ft from a reflex photoelectric sensor has about a
0.5° observation angle between the source and detector objective lenses. These narrow observation angles match well with
the performance of plastic molded corner cube reflectors. While
beaded reflectors are not as bright as corner cube reflectors,
they are visible over a wider observation angle, a property that
is not useful to reflex sensors. The acceptance angle is a measure of how far off-axis the illumination can be (or how much
the reflector can be tilted) and still maintain retroreflective properties. Figure 2.16 shows the acceptance-angle performance of

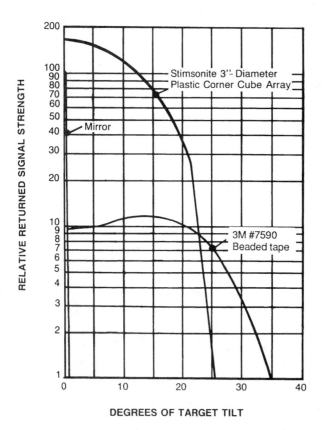

DEGREES OF TARGET TILT

Figure 2.16 The acceptance angle of retroreflectors makes them
much easier to use than mirrors.

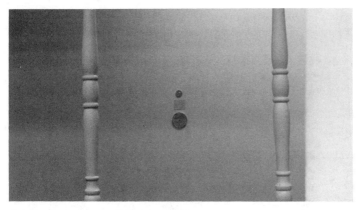

(a)

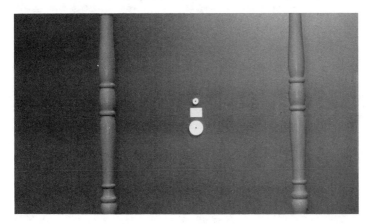

(b)

Figure 2.17 Photographs of retroreflective targets (a) in ambient light and (b) with flash illumination adjacent to the lens demonstrate the different properties of specular and retroreflectors.

beaded and corner cube retroreflectors measured with the optics of a production reflex sensor at 24 in. of range. A mirror must be aligned within about 0.5° to be a useful (or interfering) reflector, while 20 to 30° is available for carefree alignment of retroreflectors. Figure 2.17 demonstrates the difference in

reflectivity of diffuse and retroreflective materials under conditions of diffuse lighting and the near-coaxial lighting of a camera
flash attachment. The exposures were adjusted so that the
painted wall would retain the same brightness. The retroreflective tape and the corner cube targets demonstrate the retroreflective principle by their full illumination.

2.3 TRANSMISSION

Transmission is the process of conveying a desired signal from
one place to another. In the process of traveling from the light
source to the detector, the light energy may encounter dust particles, steam, chips, liquid droplets, and other forms of contamination that will scatter and absorb the light signal. Most often
we want the photoelectric sensor to operate in the presence of
this contamination and still be able to detect the object of interest. Because light transmission is affected by reflection, refraction, scattering, absorption, and polarization, understanding how they affect photoelectric sensors is key to being able to
control them so that reliable photoelectric sensing solutions can
be implemented.

2.3.1 Refraction

Refraction is the bending of light rays as they go from one optical medium into another. The index of refraction is the ratio of
the speed of light in a vacuum relative to the medium being measured:

$$\text{Index of refraction } n = \frac{\text{speed of light in vacuum}}{\text{speed of light in material}} \quad (2.5)$$

Light travels faster in a vacuum than in transparent solids, liquids, or gases. Table 2.1 lists the refractive index of several
common materials. One can show that as light slows down on en-
try into a medium of different refractive index, it bends according to Snell's law of refraction, as shown in Eq. 2.6 and Fig.
2.18.

$$\text{Snell's law: } n_1 \sin \theta_1 = n_2 \sin \theta_2 \quad (2.6)$$

In this example with plate glass, if the incident light is at an
angle of 30° from the normal (perpendicular to the surface), we
can calculate that the light will travel through the glass (with
refractive index of 1.5) at an angle 19.47° from the normal.

Table 2.1 Index of Refraction of Some Common
Materials at 520nm (Green) Wavelength

Media	Refractive index
Vacuum	1.0000
Air	1.0003
Water	1.33
Ethyl alcohol	1.36
Quartz	1.46
Acrylic	1.49
Crown glass	1.52
Sodium chloride	1.53
Polycarbonate	1.59
Styrene	1.59
Flint glass	1.66
Diamond	2.0
Silicon	3.4

Sources: The Photonics Design and Application Handbook, Book 2, Optical Plastics (Pitts-
field, Mass. Optical Publishing Co., Inc., 1985),
p. H-212; also Halliday and Resnick (1978), p.
939.

Refraction is a common sight in our everyday lives. It occurs
in fish tanks, prisms, and water glasses (see Fig. 2.19) and is
the principle that causes a lens to function. Its specific useful-
ness in fiber optics is discussed in Section 2.5. Understanding
and taking advantage of the refraction principle can be quite important to the successful application of photoelectric sensors
where glass and liquid detection are required. Figure 2.20 demonstrates how refraction alters the transmission of light through
a glass bottle when it is filled with a liquid. The difference in
the transmitted light pattern can be used to distinguish between
them. Both the light void and the focused bright spot on the

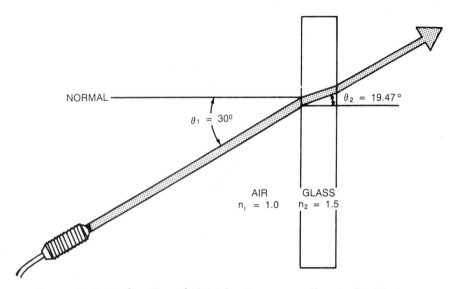

NORMAL

$\theta_1 = 30^0$

$\theta_2 = 19.47°$

AIR
$n_1 = 1.0$

GLASS
$n_2 = 1.5$

Figure 2.18 Refraction of light in glass according to Snell's law in media of different refractive index.

far side of the bottle provide significant detection opportunities, due to their extreme change in transmitted light-intensity levels.

When light goes from one refractive index medium to another, some of the light is reflected at the surface boundaries. This is called Fresnel reflection. Fresnel reflection is a transmission loss mechanism, and is the cause of the reflection glare observed on water, glass, and plastic objects. The reflection coefficient may be calculated as

$$\text{Fresnel reflection coefficient} = \frac{(n_2 - n_1)^2}{(n_2 + n_1)^2} \qquad (2.7)$$

Black tinted glass, with a refractive index of 1.5, will reflect 4% of incident light. A transparent glass plate will reflect 4% from both the front and back surfaces, for a total of 8% reflection of incident light.

A special case in refraction occurs when the angle of refraction is calculated as 90° or greater. It is not physically possible for light to exit a surface at greater than 90° from the surface normal. What actually happens is that all of the light is reflected internally. This phenomenon is called total internal reflection.

Figure 2.19 Distortion of image caused by refraction.

It can occur only when light travels toward the boundary of a medium with a lower index of refraction, as when light goes from glass to air, as shown in Fig. 2.21. The lowest angle of incidence at which total internal reflection occurs, called the critical angle, may be computed as

$$\text{Critical angle } \theta_c = \sin^{-1} \frac{n_2}{n_1} \tag{2.8}$$

Using the values in Table 2.1 and the equation for the critical angle, one can calculate that the critical angle for a water-air interface is 48.77°. A beam of light traveling 50° to the water surface normal cannot exit the water but will be totally reflected back into the body of water.

2.3.2 Scattering and Absorption

Scattering is the process by which light is redirected randomly due to reflection or refraction by many tiny particles suspended in the light-transmitting medium. Absorption is the process of materials receiving light energy and converting it into thermal

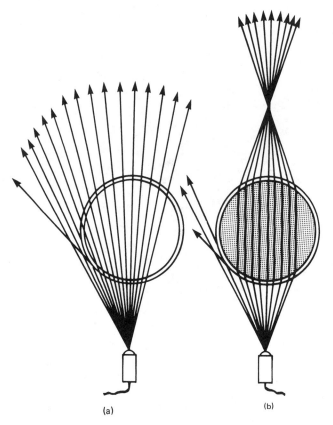

Figure 2.20 Light transmission: (a) in an empty jar; (b) through a water-filled jar.

energy rather than passing it on in the same or other directions. Both scattering and absorption reduce the amount of light available to the detector. Figure 2.22 illustrates how significant an issue scattering and absorption is in some industries. In spite of the thick sawdust on the pictured thru-beam photoelectric sensor, it still performs its detection task.

The most common contamination is the buildup of dirt or water drops on the lens. A water droplet on a lens effectively acts as an additional tiny lens on the lens. This extra little lens will defocus the light collimated by the photoelectric sensor's objective

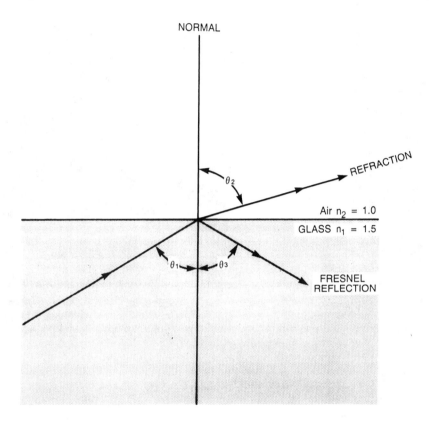

Figure 2.21 Total internal reflection occurs when θ_1 increases until θ_2 becomes greater than 90°.

lens and scatter it almost as wide as would a Lambertian diffuser (see Fig. 2.23). Generally, water and oil drops do not cover much of the lens surface before they start to join together and smooth out. The net result is that about 10% of the signal gets through anyway. However, keep in mind that the same 10% applies to the received light trying to get through the detector lens. Ten percent of 10% nets only 1% of the original signal. If we are operating with a retroreflector, this 1% factor may also apply to it in addition. As the signal goes in and out of the front surface of the retroreflector it again may be attenuated to 1% of the incident light leaving a net of only 0.01% of the original available signal strength. This would require a safety

Figure 2.22 A sawdust-covered thru-beam photoelectric control continues to operate properly in thick contamination due to extremely high excess gain characteristic of thru-beam photoelectrics.

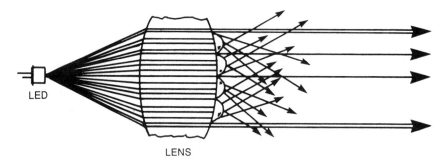

Figure 2.23 Water droplets on the lens defocus the columnated light, significantly reducing the light energy directed toward the target.

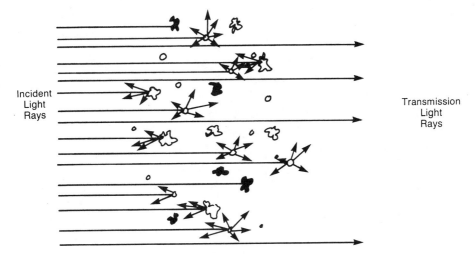

Figure 2.24 Scattering and absorption in airborne dust particles and water droplets.

margin of 10,000 times the signal strength needed in a clean, dry system! Clearly, lens cleanliness is of utmost importance when there is not a lot of performance to spare. Dirt buildup on a lens can be even more serious since it will absorb the light rather than scatter it. Again, we must determine what percent of each light-transmitting surface is free from dirt. Multiplication of these fractions together will determine what fraction of the light will be returned to the detector.

Airborne smoke particles, steam, and dust are attenuating in themselves, but are usually less of a problem than that caused by their accumulation over time on the lens or reflector surfaces. The amount of signal attenuation is related to the density of the airborne material as well as the length of travel through it. The mean free path of light through airborne material is the distance light travels before half of it has encountered a particle and been deflected or absorbed. In Fig. 2.24 the light has been attenuated by 50% in its travel. If it were to travel through the same density of smog for three times the distance, then only 12.5% of the light would be transmitted. Attenuation is an exponential function rather than a linear function of the distance light travels in an absorbing medium.

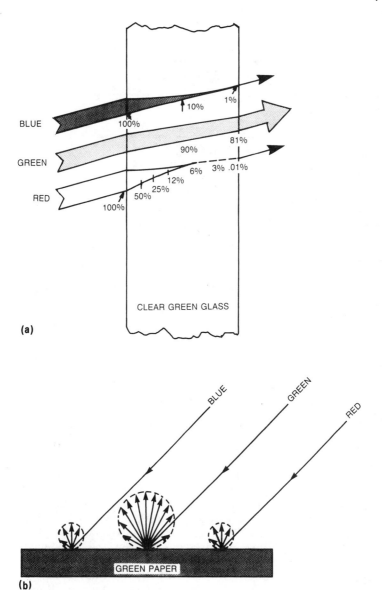

Figure 2.25 Spectral absorption of light: (a) in a transparent colored medium; (b) in a diffuse reflecting colored surface. (Courtesy of Opcon, Inc.)

Absorption and transmission can be partial to specific wavelengths creating a perceived color to the eye as depicted in Fig.
2.25. Diffuse reflecting objects usually are not as pure in color
as we would believe. It is difficult to dye or print a material
well enough to reduce any color component below about 10%. One
of the problems is that there is always some surface specular reflection that is not subject to the dye filtering. Even black paper
and flat black spray paint generally reflect more than 7% of incident light. Registration marks used by machines to keep the
printed pattern of bags, boxes, and other roll stock packaging
materials properly oriented and cut are often printed in one of
the colors present in the printed design. This often presents
quite a challenge for detection by a monochromatic LED light
source since almost any color or mix of color shades might be
used. Figure 2.26 summarizes the many previously discussed
attenuation sources light may encounter. Each of these takes
its individual toll in reducing the available light for detection.

2.3.3 Polarization

Most all of natural and man-made light sources produce unpolarized light. Light is electromagnetic radiation, that is, radiated

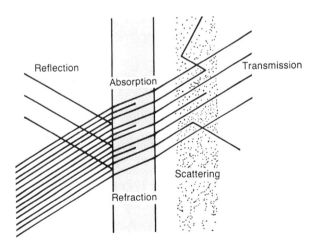

Figure 2.26 Summary of the many attenuation sources light en-
counters in the environment. (Courtesy of Opcon, Inc.)

energy with interrelated alternating electric and magnetic fields.
As such, each photon (packet of light energy) has a specific electric and magnetic field direction perpendicular to each other and
to the direction of light propagation, as shown in Fig. 2.27. The
direction of the electric field is said to be the direction of polarization of the light. The direction of polarization is generally
random. A polarizer is a material that is transparent to one direction of polarization, and opaque to polarization 90° rotated
from it. The former is said to be parallel polarized and the lat-
ter, cross polarized. Light that has a polarization direction
somewhere between these is neither parallel nor cross polarized.
Its polarization, however, may be broken down into parallel and
cross-polarized components using vector arithmetic shown in
Fig. 2.28. The parallel component will be transmitted while the
cross component will be absorbed. Light incident on a polarizer
need not be perfectly parallel polarized in order to have some
transmission. However, only the parallel polarized component
of the incident light will be transmitted. Figure 2.29 shows how
a polarizer placed at 45° between two crossed polarizers performs
the action indicated in Fig. 2.28 twice to result in a net transmission of light.

Plastic under stress, including the residual internal stress
within injection-molded plastic, has the property of rotating the
polarization of light. This property is called photoelasticity.

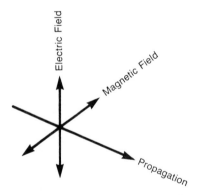

Figure 2.27 Electric and magnetic fields are both perpendicular
to the direction of light propagation. The light is polarized in
the direction of the electric field.

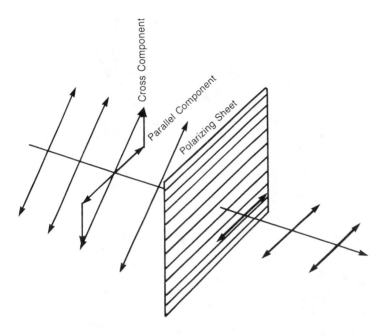

Figure 2.28 Light that is neither parallel nor cross polarized may be broken down into parallel and cross components by vector arithmetic. The parallel component will pass through the filter.

Photoelasticity was first reported in 1816 by Sir David Brewster, who found that stressed clear glass examined in polarized light exhibited colored patterns (see Dolan and Murray, 1950). Observing the stress contours of plastic with polarized light as shown in Fig. 2.30 has been a common analytical tool for structural stress analysis for most of this century. This cross-polarized test set up is called a polariscope. Injection-molded plastic generally has a great deal of internal stress associated with sharp corners due to faster cooling of the surface and later shrinkage of the interior material. Most plastics become birefringent under the application of stress. Birefringence means that the material has two different indices of refraction. The index of refraction in the direction of the stress becomes altered by the stress. The index of refraction perpendicular to the stress remains unaltered. When the incident light is neither parallel nor cross polarized relative to the direction of the stress, the light wave is divided

Figure 2.29 A polarizing material between two cross polarizers
alters the plane of polarization from filter to filter, allowing light
to be transmitted.

between the two indices of refraction and travels at two different
speeds through the plastic. When the two components of the light
wave exit the plastic, they are likely to be slightly phase shifted
from one another and recombine to form a rotating elliptically polarized wave. The rotating nature of the resultant polarized wave
will always have a component of polarization that can pass through
the second polarizing filter. Plastic corner cube reflectors with
their inherent corners everywhere would clearly be expected to
have a great deal of internal stress and impact on light polarization. A polarized photoelectric sensor and its plastic corner cube

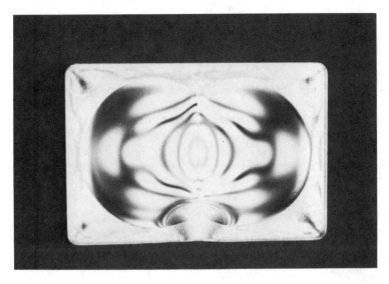

Figure 2.30 Injection-molded plastic, with residual internal stress, alters the light polarization.

retroreflector operate in the same fashion as a device that Dolan and Murray (1950) called a doubling polariscope. This device transmits monochromatic light through a first polarizer toward a target. The target is composed of a specular reflecting surface with a thin plastic coating. As the surface is stressed, so is the plastic. The returned light is observed through a cross-polarized filter. Any stress in the plastic, as in the retroreflector, will cause the specularly reflected light to be rotated and pass through the second polarizing filter. This is the property that allows polarized reflex sensors to respond to plastic corner cube retroreflectors, but not to possible specular surface reflections of would-be beam-blocking objects. Beaded retroreflectors do not alter the polarization of light and will return no signal through cross polarizers.

2.4 THE SPECTRUM

For human eyes, color is the most discriminable feature of most objects. But modern pulse-modulated photoelectrics are all monochromatic, at least at the present time. This means that they sense only how much of their own beam a surface reflects. The

effects of color can be quite deceiving in photoelectric sensing.
Most photoelectric sensors use infrared light to sense objects.
Because our eyes are not sensitive to this portion of the electromagnetic spectrum, it is difficult, if not impossible, to guess just
how the sensor will react. Even sensors using visible red or
green light do not always present us with intuitive results. As
indicated by Rock (1984), our perception of illumination levels
and the reflectivity of objects is not absolute, but is instead related to the contrast with surrounding adjacent surface brightness and color. However, a little understanding of how colored
light, objects, and detectors interact may help improve educated
guesses.

2.4.1 Source and Detector Spectra

Light sources come in many colors, as shown in Figs. 2.31 and
2.32. The sun, incandescent bulbs, and stove heating elements
emit electromagnetic energy having the classic wavelength distribution of a blackbody radiator. Thermal light sources are the
most common type in our world, have been thoroughly characterized, and are well understood by physicists today. As the temperature increases, the radiated energy increases according to
the Stefan-Boltzmann fourth-power law, the color of the peak
radiation shifts toward blue, increasing in frequency directly
with temperature according to Wien's displacement law, and the
energy is distributed in a leaning humped fashion over the spectrum according to Plank's radiation law as described by Shortley
and Williams (1971). We are all familiar with the different color
of light given off by different temperature thermal sources. Although our eye naturally adjusts to the different color temperatures of the sun and incandescent bulbs, photographic film
and electronic sensors do not. They must be color filtered to
compensate for the different spectral tilts. Incandescent light
sources radiate heavily in the infrared. In fact, direct sunlight
falling on an object to be detected may be 100,000 times brighter
than the pulsed light signal a photoelectric sensor is trying to
detect, a significant source of ambient light interference to be
dealt with. Light sources such as fluorescent lamps, mercury
vapor lamps, and high-pressure sodium lamps produce light almost entirely in the ultraviolet and visible spectrum. These AC
line-powered lamps produce light modulated at 120 Hz (twice 60
cycles per second) and harmonic multiples of 120 Hz. Pulsating
light produces the stroboscopic effect we see when fast-moving
objects such as wheels and fans are observed in street lighting.

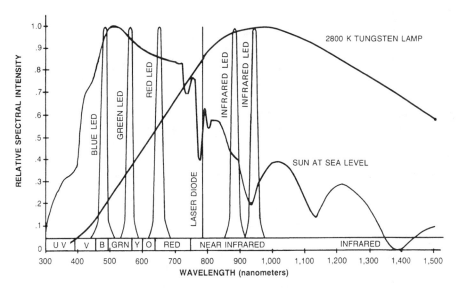

Figure 2.31 Spectral emission of common thermal and solid-state light sources. [From *IES Lighting Handbook*, 5th ed. (New York: Illuminating Engineering Society, 1978), pp. 8-17 to 8-20, 25-1.]

Pulsed photoelectric sensors must distinguish between an overhead lamp's pulsating light and its own. Infrared LED photoelectric sensors will often employ a very dark red plastic filter in front of or behind the lens with the characteristics shown in Fig. 2.32. The infrared pass filter virtually eliminates all visible-light interference before it gets to the detector--a significant help in reducing interference from these lamps.

Solid-state lamps emit light in very narrowly defined spectra, depending on the material from which they are made. Infrared LEDs have become the dominant light source used in photoelectric sensors due to their long life, ability to be pulsed at high frequencies, high efficiency, and high coupling efficiency with silicon photodiodes. A silicon photodiode has a much wider spectral sensitivity than does the human eye. Silicon is most sensitive in the infrared and fair to poor in the blue and near ultraviolet, as

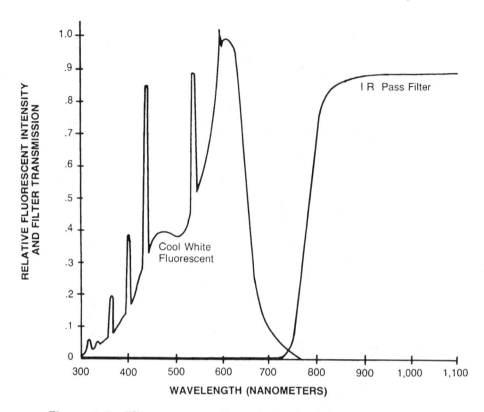

Figure 2.32 Filters are sometimes used with infrared sensors to reduce interference from fluorescent and other visible gas discharge lamps. [From *IES Lighting Handbook*, 5th ed. (New York: Illuminating Engineering Society, 1978), pp. 8-17 to 8-20, 25-1; *Kodak Filters for Scientific and Technical Use*, catalog number 152-8108 (Rochester, N.Y., 1970), Eastman Kodak Company, pp. 8, 69--79.]

shown in Fig. 2.33. Silicon has the advantage of being able to detect light that our eyes cannot. However, this advantage carries the liability of causing human beings to be poor judges in estimating performance of these sensors. Visible LEDs have become more popular as their performance has increased and can be useful in applications where the sensing of color is important.

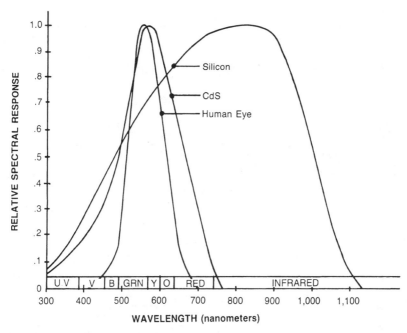

Figure 2.33 Spectral response of common light detectors. [From *RCA Electro-Optics Handbook*, Technical Series EOH-11 (Harrison, N.J.): RCA Corporation, pp. 56, 62, 79, 160.]

2.4.2 Filters and Dyes

Color is particularly important in proximity sensing of printed registration marks on packaging materials or when one must detect transparent colored liquids and films. Many organic-based dyes attenuate well in various parts of the visible spectrum but become transparent in the infrared. This is not immediately apparent to our infrared-insensitive eyes. Kodak Wratten filters, for example, are produced using dissolved organic dyes in a gelatin base. Blue filter number 35 and green filter number 57, shown in Fig. 2.34, demonstrate that infrared light at 880 nm would be a poor choice of illumination for detection of these materials. Generally, inks with carbon black or dark blue dyes are the easiest to detect with any LED color. Successful presence/absence detection relies on the sensor's ability to obtain sufficient contrast to make the decision. Contrast and color are discussed in detail in Section 2.7.

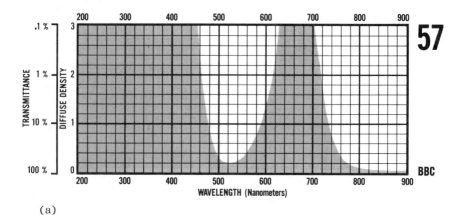

(a)

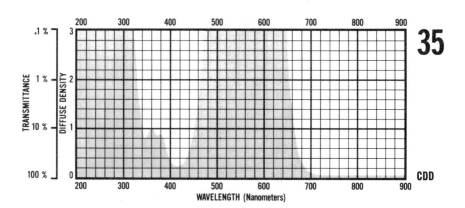

(b)

Figure 2.34 Kodak Wratten filters (a) number 57 green and (b) number 35 blue demonstrate how poorly infrared light would act as a sensor for these typical organic dyes. (Reprinted courtesy of Eastman Kodak Company.)

Fiber optic materials also have color-dependent attenuation properties, particularly in plastic fibers. Figure 2.35 shows spectral attenuation curves for glass fiber bundles and plastic fibers typically used with photoelectric sensors. The attenuation of infrared light in plastic fibers makes them particularly unsuitable for long-distance fiber runs. To make use of these

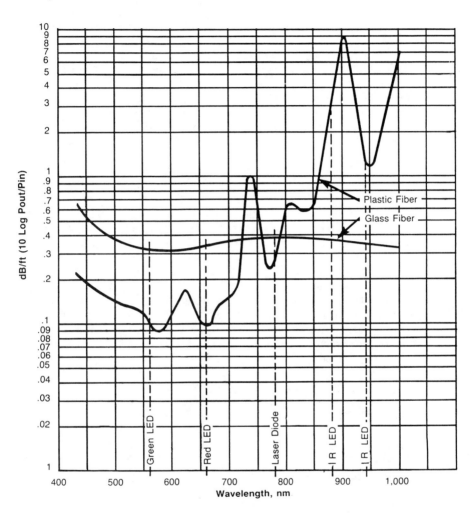

Figure 2.35 Attenuation of light in glass and plastic fiber optic cables typically used in photoelectric sensors.

graphs, one needs to know three basic pieces of information. First, determine what color or wavelength is being used. Second, determine the total distance light will travel in the fiber (out and back). Third, each 3 dB of attenuation represents a

50% reduction in signal strength. Let's take the example of the plastic fiber running to a sense point 15 ft away (30 ft round trip). A red LED at 660 nm will be attenuated by 0.1 dB/ft for a total of 3 dB. Thus the visible LED will lose 50% of its optical strength in the fiber. An infrared LED at 880 nm will be attenuated at 3 dB/ft for a total loss of 90 dB or 99.9999999%, or alternatively, 0.0000001% transmission. The transmission percentage may be calculated directly using the formula

$$\text{Transmission } \% = 10^{2-(\text{dB loss/ft})(\text{length})(0.1)} \qquad (2.9)$$

2.4.3 Fluorescence

Fluorescence is the reemission of absorbed light from a substance at wavelengths longer than that of the incident stimulating light. Fluorescent lamp tubes are the most common example. In these tubes, an arc of current flowing through the mercury vapor within the tube excites the mercury atoms, causing them to radiate ultraviolet (UV) light at a wavelength of 254 nm. When the UV light strikes the phosphor coating on the inside wall of the tube, it excites the electrons of the phosphor atoms temporarily into a higher energy state. As the electrons return to their normal energy states, visible light is emitted.

Fluorescence has two practical uses with photoelectric sensors. First, an invisible detection mark of some sort is occasionally required when a visible spot or stripe would be aesthetically displeasing. In these cases it is sometimes possible to use a transparent UV fluorescent ink. The ink is illuminated by a UV source as it passes the sensor. The fluorescent ink absorbs the UV light and reradiates it in the visible part of the spectrum where a silicon photodiode can detect it. Special photoelectric sensors have been designed for this purpose. Second, Eastman Kodak Company produces an IR (infrared) phosphor that requires a double stimulation for emission. The IR phosphor makes it possible to view the infrared radiation of a photoelectric infrared LED source. The phosphor is first activated by blue visible light from a fluorescent lamp. The blue light excites the phosphor electrons into a stable higher energy state. When exposed to infrared light, the electrons are stimulated to fall back to their lower energy state and emit an orange light. Some photoelectric manufacturers have IR phosphor-coated test cards available as an accessory.

2.5 FIBER OPTICS

The fact that light could be transmitted around a corner was first demonstrated by John Tyndall to the British Royal Society in 1870 as he caused a light to shine along a stream of water flowing from a hole in the side of a tank. As the stream arced toward the floor, the light arced with it; remaining in the stream of water rather than departing from it as the water changed directions. The light was captivated in the stream by total internal reflection, described in Section 2.3, at the water-air boundary. This is the principle of glass and plastic fiber optic operation. The utility fiber optics brings to sensing is tremendous and includes small sensing geometry, high-voltage isolation, very high temperature operation, and nonunion installations.

2.5.1 Total Internal Reflection

For light transmission, the fiber optic cable is analogous to the transmission of electric current in wire. In the case of wire, electric current is well confined by the boundaries of the wire conductor. Neither air nor plastic insulation is a conductor or absorber of electric current. However, light is easily transmitted by air and absorbed by materials that might act as a protective jacket to an optical fiber. Therefore, an optical fiber cable must be constructed in a manner that prevents the internally traveling light from having any interaction with surrounding air or jacket materials. This is accomplished by causing total internal reflection to occur before the light reaches the surface of the fiber. This is done by cladding the fiber core with another transparent material of lower refractive index, as shown in Fig. 2.36. Within the clad fiber, total internal reflection will occur at the core-to-clad interface as long as the angle of incidence is greater than the critical angle measured relative to the boundary surface normal (a line perpendicular to that surface). As an example, assume that a stepped-index fiber, as in Fig. 2.36, were constructed with a core refractive index of $n_2 = 1.50$ and a cladding refractive index of $n_1 = 1.40$. The critical angle, as calculated from Eq. 2.8, would be 69° from normal, or 21° off the fiber axis. Refraction at the fiber ends will cause a 21° off-axis internal light ray to be bent further, to 32.6° off-axis, as it enters the air according to Snell's law (Eq. 2.6) and as shown in Fig. 2.37. Light incident on the fiber core more than 32.6° off-axis will not have total internal reflection, but instead will enter

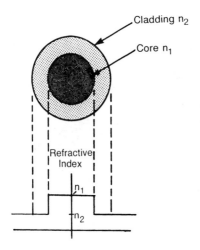

Figure 2.36 Stepped-index fibers have a core with high refractive index surrounded by cladding with a lower refractive index.

the cladding and likely be absorbed by the fiber jacket. Light incident on the fiber core at less than 32.6° off-axis will have total internal reflection and will be transmitted along the fiber core. The acceptance angle for off-axis light for this fiber is 32.6°. The acceptance angle may be calculated directly using Eq. 2.10 as shown by Seippel (1981);

$$\text{acceptance angle } \theta_a = \sin^{-1}(n_2{}^2 - n_1{}^2)^{1/2} \qquad (2.10)$$

Often this information will be given in terms of the numerical aperture (N.A.), which is closely related to the acceptance angle:

$$\text{numerical aperture (N.A.)} = \sin \theta_a = (n_2{}^2 - n_1{}^2)^{1/2} \qquad (2.11)$$

The acceptance angle, in three dimensions, is really an acceptance cone. Two rays entering the acceptance cone at different angles will travel different lengths through the fiber, as depicted in Fig. 2.38. Rays that travel directly down the center of the fiber will emerge first since their path is shorter than alternative zigzag paths. The difference in transmission time is called dispersion. Dispersion is important in telecommunications

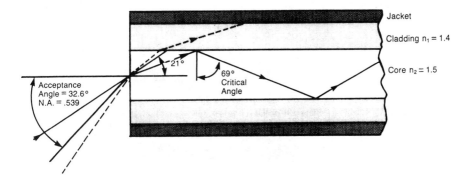

Figure 2.37 The acceptance angle is the largest angle at which an incident light ray may enter the fiber core and be transmitted to the other end via total internal reflection.

because it limits the rate at which information may be transmitted by smearing light pulses into one another and is inversely proportional to the fiber length. Graded-index fibers (Fig. 2.39) are used to decrease dispersion by causing rays traveling straight down the center to travel slower, and those that zigzag, to travel faster as they get near the cladding where the refractive index is less. Graded fibers are of no advantage in photoelectric sensors since it would require at least 50 miles of fibers for time dispersion to affect significantly the pulses of a typical photoelectric sensor.

2.5.2 Glass Versus Plastic Fibers

There are two principal types of fiber optic cables used industrially with photoelectric controls: the multiple glass fiber bundle, and the single plastic monofiber. The glass fiber bundle has the advantage of withstanding temperatures as high as 451°F (232°C) as standard product [see Bradbury (1953) for additional discussion], and is generally manufacturable on special order at rated temperatures up to 900°F (482°C). Hundreds of glass fibers are embedded in a high-temperature epoxy in the bundle tip. The tip is then well polished to ensure that each fiber tip is faceted

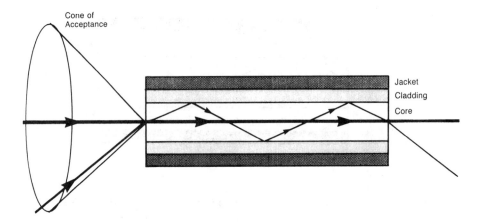

Figure 2.38 The distance traveled by an off-axis ray is longer due to its zigzag path.

uniformly. The need to factory cut and polish glass fiber bundles requires that nonstandard cable lengths be special-ordered. Glass fibers are usually jacketed with an interlocking stainless steel armor or polyvinyl chloride (PVC) jacket.

The PVC-coated plastic monofiber is rated at temperatures up to 175°F (80°C). They are generally available precut with integral mounting threads, but may also easily be cut to length in the field. The low cost, small size, and ease of field trimming plastic fibers presents benefit trade-offs that are quite attractive in some instances.

Both glass and plastic fibers have physical stress limitations. Glass fibers are easily broken by pulling, stretching, twisting, and extreme vibration. Glass fibers are capable of withstanding indefinite cyclic bending with reasonable radius corners. Sharp bends, rubbing, or pull stress is usually not equally distributed among the fibers. Individual fibers can easily receive inordinate stress and fracture. Breakage of a few fibers is not generally detectable. However, fiber breakage affects signal strength proportionately and will become a problem if significant numbers do break. Although these cables have a tough and flexible physical feel, they should be treated as if glass were packaged inside.

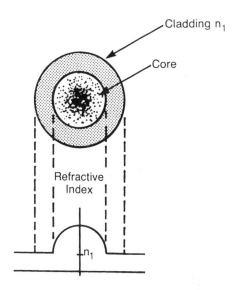

Figure 2.39 Graded-index fibers have a core with refractive index that decreases as the distance from the center increases.

Plastic fibers are not generally susceptible to breakage. Because of the physical size of the individual fiber, a sharp corner will impact the angle at which light strikes the core-clad boundary. Light going straight down the center of the fiber will stray into the cladding boundary at the bend and be reflected back toward the fiber center. However, light already traveling near the critical angle may go beyond the critical angle at the bend, enter the cladding, and be absorbed by the jacket. Losses at corners can be virtually eliminated by keeping the corner radius greater than 10 times the fiber diameter. Plastic fibers are also susceptible to fogging with excessive stress. Fogging is the result of stress-caused microcracking in the plastic structure. This causes attenuation by scattering. Plastic fibers are susceptible to the same stress sources as those of glass fibers. In either case, these fibers are suitable for industrial use, but must be treated with respect.

Coherent fiber bundles keep the physical order of fibers the same at both ends of the cable. Coherent fiber bundles are quite expensive to manufacture and are most commonly used in medical instruments and borescopes to transmit an image from one end of

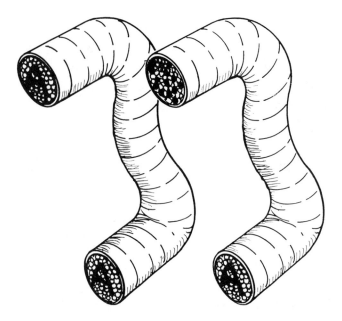

Figure 2.40 Coherent fiber bundles transmit an image from one
end to the other. Randomized fiber bundles scramble the image.

the bundle to the other. Figure 2.40 shows how an image of the
letter "A" is transmitted to the opposite end of the coherent bundle, but is converted into little black spots of dark fibers in a
randomized bundle. Photoelectric sensors commonly use inexpensive randomized fiber bundles. The quantity of light energy incident on the bundle tip is more important than the light energy
pattern.

Often the fiber cables for the source and detector are combined in a "Y" junction as in Fig. 2.41 to form a single bundle
tip at the sense point. This split cable is called a bifurcated
cable. It has significant advantages in mounting and signal coupling for close proximity sensing.

2.5.3 Spectral Attenuation

Attenuation of light as it travels through optical fibers depends
on the color (or wavelength) of the light, the fiber material, and

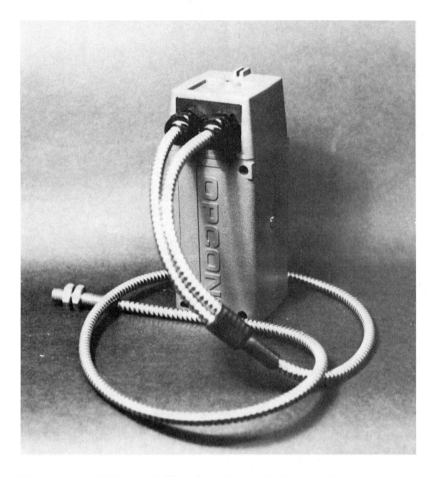

Figure 2.41 Bifurcated fiber bundles split the bundle from a single sense point to couple light out from the source and back to the detector.

the distance traveled in the fiber. Glass fibers are quite uniform in performance at all wavelengths used in photoelectric sensors (see Fig. 2.35). There are much-lower-loss glass fibers available that cater to the telecommunications market. However, they are much more expensive than the utility grade typically used in the photoelectric market and would give insignificant additional

benefit for the 3-ft-long cables generally available as standard
product. Plastic fibers attenuate very little in the visible and
quite heavily in the infrared wavelengths. This does not mean
that infrared light has no relevant value with plastic fibers. It
means that infrared light is restricted for use at shorter distances with plastic fibers. Let's look at this situation more
closely. Infrared LEDs rated at 880 nm, are widely used in
photoelectric sensors because of their high light output and
spectral match to the silicon photodiode sensitivity peak. In
fact, infrared LED sensors are typically more than 10 times
more powerful than their visible red beam counterparts. Referring to Fig. 2.42, we observe that at 1.7 ft, only 10% of
light is transmitted in the plastic fibers at 880 nm. However,
this just reduces its optical strength down to the equivalent

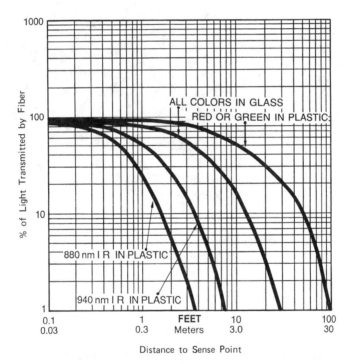

Figure 2.42 Attenuation of light in an optical fiber depends on
the fiber material, color of light, and distance traveled in the
fiber.

level of their red LED counterparts. At 10 ft, the red and
green units will be at about 60% of rated performance, but the
880-nm infrared units will be at a useless 0.0001% of rated performance. As discussed in Sections 2.4 and 2.7, there are often
good reasons why a particular color of light should be used for
an application. Also, fiber optic cable need not always be a
limiting factor. Because there are many grades of optical fibers, applications assistance should be requested for exact performance at cable lengths or wavelengths not specified by the
manufacturer's catalog.

2.5.4 End Termination

Coupled signal strength in fiber optic sensors also depends on
the size of the fiber bundle tip. A bundle tip diameter of 0.0625
in. will have only one-fourth of the light-collecting area as a
bundle with a diameter of 0.125 in. There are two places that
light enters the fiber cable in a photoelectric sensor. Light from
the LED must enter the source fiber cable, and light returning
from the target must enter the detector fiber cable. This results
in a coupling reduction of 1/4 × 1/4, or 1/16, of the signal expected from fiber cables with twice the diameter. The coupled
signal strength can be increased by the addition of a lens mounted on the fiber tip. The lens increases the light-collecting area
of the detector cable, and increases the light intensity from the
source cable in the far field. A lensed tip source cable does not
magically increase the available light; rather, it redirects light
exiting the cable at wide angles into a more collimated beam.
The lens provides the same functional advantage for the fiber
cable as it does for photoelements on a photoelectric sensor.

2.6 EXCESS GAIN

Excess gain is an overall performance specification for photoelectric sensors and controls. Used properly, excess gain figures
can tell you how much deviation from best-case optical operating
conditions can be tolerated and can aid in selection of the best
sensor for the task.

2.6.1 Definition

Excess gain is a measure of the extra signal energy available to
a photoelectric sensor for overcoming attenuation in the beam

path. Excess gain is the operating margin available for "seeing" through dirt, steam, liquids, and translucent objects. It improves performance with small tipped fiber optics and long-range sensing. Very dark objects require high excess gain for diffuse proximity sensing. Excess gain (G_X) is defined as the ratio of signal level received at the detector to the threshold signal level required for detection:

$$G_X = \frac{\text{signal level received}}{\text{detector threshold level}} \qquad (2.12)$$

For $G_X = 1.0$, only the minimum signal level required for detection is present. For $G_X = 15$, the received signal level is 15 times stronger than the threshold level. Excess gain is normally specified as a function of range on a log-log scale plot as shown in Fig. 2.43. Excess gain is particularly important for estimating performance in contaminated environments. Excess gain should not be confused with range of operation! Although there is a relationship between excess gain and range, the shapes and slopes of the excess gain curves vary considerably among sensor designs and types. If the intent is to operate at a 1-ft range, then

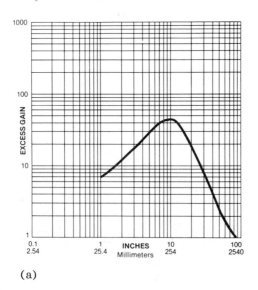

(a)

Figure 2.43 Typical excess gain curves, (a) proximity, (b) reflex, and (c) thru-beam, graphed on log-log scaled paper. (Courtesy of Opcon, Inc.)

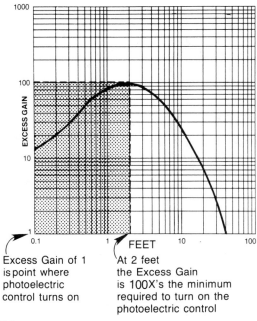

Excess Gain of 1
is point where
photoelectric
control turns on

At 2 feet
the Excess Gain
is 100X's the minimum
required to turn on the
photoelectric control

(b)

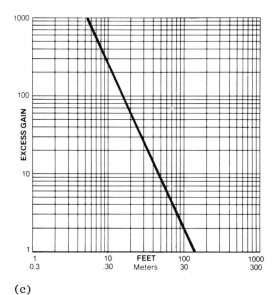

(c)

excess gain performance of photoelectrics should be compared at
the 1-ft range. This is particularly important with proximity
sensors which have significant variations in optical design in
order to optimize performance in specific operating ranges.

Net operating margin is a closely related and important con-
cept. Net operating margin takes into account the effects of the
target. For example, not all proximity targets are white diffuse
targets. If we had a black target that was only 10% reflective
relative to standard white and a sensor that had $G_X = 15$ at the
detection range, the black target would return $15 \times 10\% = 1.5$
times the signal required to reach detection threshold. We would
then say that the net operating margin for this application is 1.5.
Net operating margin tells us what safety factor remains to ensure the reliability of detection.

2.6.2 Sensor Characteristics

Excess gain is always specified for a specific standard target.
Reflex sensors will specify the type and size of retroreflector.
Diffuse proximity sensors are specified relative to a standard 90%
reflectance Kodak test card. Thru-beam sensors are specified
without obstruction between them.

In diffuse proximity sensing, objects we wish to detect are
often better or poorer reflectors than we expect. Sometimes they
are much better because the have a partial specular peak that
can be used to advantage. Often, the infrared or monochromatic
light of the sensor causes the object to appear differently to the
sensor than to our eyes. Table 2.2 provides a few typical reflectivity values for some common objects relative to the Kodak
test card for 880-nm infrared light most commonly used in infrared photoelectrics. In the right-hand column is the minimum
recommended excess gain for sensing these objects, including
the safety margin of 1.5× used in a relatively clean environment. For example, if we were required to sense black wool cloth,
Table 2.2 recommends an excess gain of at least 9. Looking at
the proximity excess gain curve of Fig. 2.43, the range over
which reliable detection is obtained is from 1.5 to 28 in. The
characteristic excess gain curve shape of proximity sensors depends strongly on the arrangement of the photoelements and
lenses. As a consequence, the only rule of thumb for estimating
proximity excess gain performance is that there is no rule of
thumb. Each sensor's specification must be evaluated separately.

In reflex sensing, the characteristic excess gain curve shape
is fairly consistent among all manufacturers' products. The reason

Table 2.2 Reflectivity of Common Objects at 880 nm
Relative to a 90% Reflecting Kodak Test Card

Material	% reflectivity	G_X recommended[a]
Lampblack	1.6	95
Black neoprene	4.4	35
Asphalt	8.5	18
Black wool cloth	18	9
Black cotton cloth	22	7
Black felt	25	6
White 0.5-mil poly bag	48	3
Plywood	60	3
Brown kraft paper	70	2
White plastic dish	75	2
White bond paper	94	1.6

[a]Recommended G_X includes 1.5× safety margin.

for this lies in the fact that the constraints in design are such that there is only one arrangement of photoelements and lenses that produce a high quality reflex sensor. The range where excess gain peaks will vary from about 1 ft to 6 ft, depending on the lens size. Overall excess gain will vary from design to design depending on the components used. However, the general shape is fixed by the required geometry of the optical components in the design. The exact performance will also depend on the size, shape, and construction of the retroreflector used. Most reflex sensors are specified using a 3-in.-diameter plastic molded corner cube retroreflector.

Only with thru-beam sensors can one count on a simple rule of thumb to relate excess gain to range. Excess gain is related to range by the inverse-square law. Using Eq. 2.13 and setting the range value equal to maximum range, the expression yields $G_X = 1$:

$$\text{Thru-beam } G_X = \frac{(\text{maximum range})^2}{(\text{range})^2} \qquad (2.13)$$

If the maximum range is 160 ft, the excess gain at 5 ft may be calculated as

$$G_X = \left(\frac{160}{5}\right)^2 = 1024$$

This relationship produces the simple straight line for a log-log excess gain plot as shown in Fig. 2.43. This typical thru-beam performance shows that excess gain exceeds 1000 at 5 ft, and in fact would exceed 10,000 at a 1-ft range. Thru-beam sensors not only can operate in heavy contamination, but are also quite useful for detecting contents inside closed packages made of paper or Mylar.

2.6.3 Environmental Effects

Consideration of the operating environment is important when selecting a photoelectric sensor. High performance can be obtained even in dirty environments if the cleanliness of the lens is maintained. Contamination buildup on the lens from settling airborne particulate matter is the worst of all contamination. Maintenance suggestions in Chapter 7 may be helpful for retaining long-term performance. Most manufacturers will recommend that in the cleanest of environments, the minimum G_X be at least 1.5. It is poor practice to operate a sensor on the marginal edge of performance when reliability is important. Table 2.3 provides a guide to excess gain required at a variety of contamination levels. These descriptions are of arm-waving quality at best and are a poor substitute for actual measurements or direct experience in the environment in question. The best performer in heavy contamination is a thru-beam system because it has extremely high excess gain at moderate ranges and has only its two lens surfaces on which contamination can collect. In addition to contamination effects, many, but not all, photoelectrics are poorly temperature compensated and may vary by as much as a factor of 2 in sensitivity over the specified operating temperature range. The most significant contributor to this problem is the LED power output, which decreases with increasing temperature.

Table 2.3 Estimating Excess Gain Required
by Environmental Considerations

Condition		Detection Method		
		Thru-Beam*	Proximity	Reflex
Relatively Clean (office buildings)		1.25 for each side (1.6)	1.6	1.6 for each side (2.6)
Lightly Dirty (warehouses, post offices, clean processes)		1.8 for each side (3.2)	3.2	3.2 for each side (10.5)
Dirty (steel mills saw mills, paper plants)		8 for each side (64)	64	64 for each side
Very Dirty (steam tunnels, car washes, rubber or carbon grinding, cutting machines w/coolants)		25 for each side (625)	— — — —	— — — —
Extremely Dirty (coal bins or where thick layers build quickly)		100 for each side (10,000)	— — — —	— — — —

(sum of a simple source and detector system in a common environment)
*Also separately located proximity detection

Source: Opcon, Inc.

2.6.4 Sensor and Target Variation

Many parts are used in the manufacture of photoelectric sensors.
Each has a specification tolerance that results in some variation
in performance from unit to unit. Excess gain curves generally,
but not always, represent the minimum performance to be expected. However, to avoid trouble with background or fore-
ground objects, it is sometimes important to know the range of
performance. It is typical to find unit-to-unit excess gain variation that exceeds a factor of 3. The performance variation range
is sometimes specified as shown in Fig. 2.44 to help with product

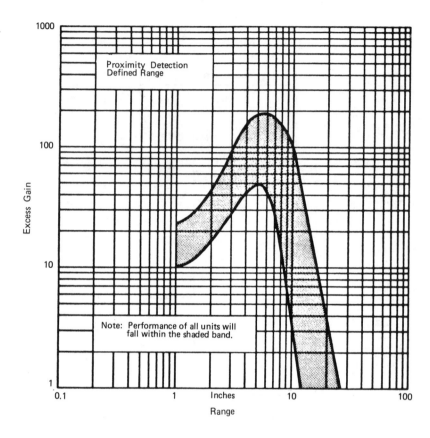

Figure 2.44 Minimum/maximum performance range specified for
a defined range proximity sensor. (Courtesy of Opcon, Inc.)

application. Most photoelectric sensors have a gain adjustment
for tailoring performance to the application requirements. Although high excess gain can be important for sensing through
industrial dirt, more is not always better. Too much excess
gain could, for example, cause a thru-beam sensor to "see"
right through the very object it is expected to detect. Contrast
ratio, described in Section 2.7, is sometimes more important than
high excess gain.

2.7 CONTRAST RATIO

It is often important to know the difference in received signal strength between the presence and absence of an object or mark. This measure, expressed as the ratio of the two signal strengths, is called "contrast ratio":

$$\text{Contrast ratio} = \frac{\text{light signal level}}{\text{dark signal level}} \qquad (2.14)$$

Contrast ratio is particularly important for reliable proximity sensing and thru-beam detection of semitransparent objects. Contrast may be the result of objects at different ranges returning different amounts of light. Contrast may come from objects or parts of objects having different colors or reflectivities. Contrast may be due to the difference in light transmission through an object. Contrast may also be related to the size of an object and how much of the effective beam it reflects or blocks.

There are some simple methods for taking the guesswork out of selecting photoelectrics and predicting performance without a lot of costly trial and error, even in the tricky applications. In the following sections there are some trial calculations, just to give you a sense of how to use data sheet values and graphs. Also, there is a simple method explained for making some optical measurements with photoelectrics themselves. With these techniques, you can get a good idea of how much margin for error you have in your application for a particular sensor.

2.7.1 Low-Contrast Targets

Let's say, for example, that our application is to detect a box on a conveyor. We can calculate the contrast ratio by using the excess gain graph for the sensor, knowing the object range, and estimating or knowing the object reflectivity. This box will be at a range of 10 in. and has a 70% reflective surface kraft paper. There is a wall in the background at a range of 30 in. with brown 40% reflective paint. The proximity sensor we wish to use has excess gain performance, as shown in Fig. 2.43. At a range of 10 in., $G_X = 45$, and at 30 in., $G_X = 7.5$. On first inspection, it seems that this sensor will see both the box and the wall and may not be useful. Often, too much excess gain can be a problem for sensing dark objects, through objects, or too far. However, being educated photoelectric consumers, we decide to calculate the contrast ratio. Since the box and the wall are not perfect reflectors, we must multiply the excess gain by the object reflectivity to get the net operating margin. The ratio of the net operating margins is the contrast ratio. By plugging in the appropriate numbers, we can calculate the contrast ratio:

$$\text{Contrast ratio} = \frac{45 \times 0.70}{7.5 \times 0.40} = 10.5$$

A contrast ratio of 10.5 is considered excellent. By adjusting the sensitivity of the sensor downward, we can set the threshold for detection between these two signal levels and reliably detect the box after all. The optimum threshold would be set at the geometric mean of the net operating margins:

$$\text{Set threshold} = [(45 \times 0.70) \times (7.5 \times 0.40)]^{1/2} = 9.7$$

The threshold should be set 9.7 times higher than the full excess gain threshold, which therefore requires that the unit be reduced in sensitivity by a factor of 9.7. Our excess gain curve is marked in Fig. 2.45 to show the net operating margins for the box and wall, and where the optimum new threshold for detection is to be set by adjustment of the sensitivity or gain control. Unfortunately, low-cost sensors do not come with dial indicators on their sensitivity adjustments and few end users have equipment for accurate measurement of surface reflectivity. However, this process can usually be accomplished successfully by estimating contrast ratio with excess gain charts, tables of reflectivity, and throwing in a little experimentation. As a practical matter for sensitivity adjustment, one simple method that approximates the optimum is to find the limits of adjustment where detection fails and then position the adjustment halfway between these limits.

2.7.2 Low-Contrast Sensor Techniques

Good contrast ratio is the key to reliable operation. How much contrast ratio is needed for reliable operation is dependent on how consistent things are. If object reflectivity, range, and lens cleanliness were perfectly consistent, a contrast ratio as low as 2.0 could be made reliable for a garden-variety sensor. Real factory conditions are not generally this nice. Table 2.4 offers some guidelines for reliable detection at different contrast levels. There are three dominant methods used for signal detection that are worthy of discussion: fixed threshold, ac-coupled rate of change, and differential.

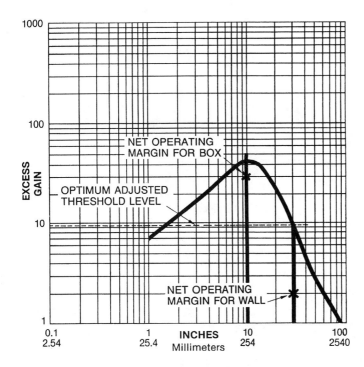

Figure 2.45 Threshold should be set midway between the net operating margins of the two objects.

Most photoelectric sensors use the fixed threshold method of detection. If the signal is above the threshold, we say that it has detected the signal. If the signal is below the threshold, there is no detection. Threshold is limited by interference noise from local electrical and optical sources that may confuse the detector if the threshold is set too low. A fixed threshold detector has no way to compensate automatically for changes in lens cleanliness, material reflectivity variations from batch to batch, or other slowly changing ambient conditions. Due to its intolerance to these external conditions, a fixed threshold sensor is generally incapable of reliable operation in low-contrast applications.

To sense low-contrast marks or objects reliably, it is necessary for the sensor to compensate for the slowly changing variables

Table 2.4 Guidelines for Contrast Ratio

Contrast ratio	Sensing recommendation
1—1.2	Unreliable: There is not enough contrast to reliably avoid false detection in most applications. Alternative sensing methods should be evaluated.
1.2—2	Poor: Consider only sensors specifically designed for low-contrast detection.
2—4	Fair: Low-contrast sensors would be best. General photoelectrics may be used only if conditions are perfectly clean and the product to be detected is of high optical consistency.
4—10	Good: Small changes in product consistency or distance will probably be tolerable.
10 and up	Excellent: There should be no problem with reliable sensing as long as contamination and limited excess gain are not issues.

mentioned above. One way this is accomplished is by ac-coupling the signal to the detector, as shown in Fig. 2.46. When circuits are ac-coupled, rapidly changing signal levels get through to the detector while the average returned signal level is rejected. Such a sensor can be made quite sensitive to low-contrast changes on rapidly moving objects. It is important to remember that the rate of change in signal level is the foundation for operation. These sensors will not detect a slow or stationary object, even with large contrasts. However, these sensors have pitfalls as well. Since signal level usually varies with distance, it is not difficult to cause a false output from a web that flutters up and down as it travels past the sensor.

The third means for low-contrast detection is differential detection. Differential detection uses the principle of comparison between two adjacent locations. The returned signal from one location is subtracted from the other so that only their difference is amplified. Figure 2.47 shows how differential sensing

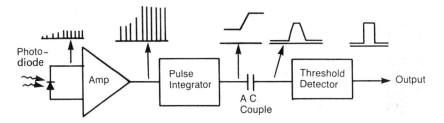

Figure 2.46 Ac coupling the detector allows low-contrast signal detection by removing steady background signals.

is not subject to speed variation problems like the ac-coupled method. In fact, differential sensing can be used on stationary targets. Its disadvantage is that it requires more hardware to implement. Some remote photoelectric sensing heads may be directly connected as shown in Fig. 2.47 to a single amplifier to perform this function. The two optical sensors are gain balanced in set up to produce a null from the differential amplifier when a uniform target is presented.

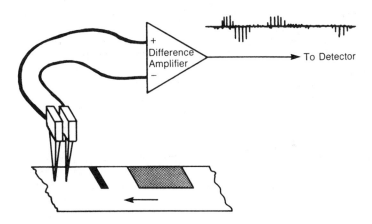

Figure 2.47 A differential sensor uses two detectors to sample nearby areas for comparison. Only light levels that are not the same are amplified for detection.

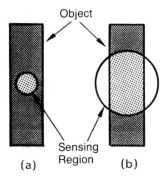

Figure 2.48 Contrast is reduced (a) when the object is smaller than the size of the effective beam.

2.7.3 Contrast and Target Size

Contrast ratio is dependent not only on reflectivity or trans-missivity ratios, but also on the relative size of the target and sensing beam. Figure 2.48b shows how an effective beam that is larger than the target can reduce the contrast ratio. An object that fails to block or encompass the full sensing region will not produce the maximum available change in signal. A comparison of the specified effective beam at the intended de-tection range with the size of the target is required to deter-mine if the expected contrast ratio can in fact be realized. In Fig. 2.48b, about 50% of the effective beam extends outside the mark. We may have been clever enough to get a paper-to-mark contrast ratio of 8:1 in ideal conditions, but our sensor choice or mark size choice could defeat us. Looking back at Eq. 2.14, for contrast ratio, let's see what happens to the contrast ratio when the mark takes up only 50% of the effective beam sens-ing area. We can calculate the dark signal level with the addi-tion of returned signal from the area in which the effective beam spills over the mark. Using a signal level of 8.0 for paper and 1.0 for the mark, we calculate:

$$\text{Contrast ratio} = \frac{8.0}{(0.5)(8.0) + (0.5)(1.0)} = 1.78$$

What was once a good contrast ratio has been reduced to a poor one. Mark or object size should, almost without exception, be

larger than the effective beam in order to obtain expected results.

2.7.4 Color Contrast

Color is a fascinating and complex subject. Human beings are
capable of incredibly sophisticated perceptions and judgments related to color sensation. Machines are far from having the same
range of capabilities, but there are a few photoelectric products
that are useful for detecting color marks on moving materials (see
Section 3.1.6), sorting by color markings, and so on. In doing
so, they are not really sensing color per se. Instead, they simply sense that the color mark reflects or absorbs a different
amount of the photoelectric's light beam than the surrounding
material. In 1965, the International Commission on Illumination
(CIE) published the report "Methods of Measuring and Specifying Color Rendering Properties of Light Sources" in order to
standardize definitions, test conditions, and measurement methods of color. Our perception of color is closely coupled to the
spectral content of light illuminating an object. Similarly, measured contrast is directly related to the light colors chosen.
Figure 2.49 shows rainbow-colored CIE chromaticity diagrams
photographed in red and green light, respectively. The color
version of the chromaticity diagram has blue in the lower left,
red in the lower right, green toward the top, and white in the
center. Clearly, red light would be a poor choice for finding a
red mark on white paper, but green light would work quite well.
It is hard to know how well infrared light would sense a red
mark since many pigments are surprisingly transparent to infrared light, although they might look quite dark to us. (Of
course, the majority of pigments were developed without any
consideration of what they do to infrared light.) The only object color that gives good contrast at all wavelengths of interest
is a carbon-based black. Table 2.5 shows the measured contrast
ratio between white and the specified printed color at red, green,
and infrared wavelengths. Some mark detectors provide more
than one LED color so that contrast may be optimized without the
need to change sensors when the same production line is changed
over to run another product. In all contrast detection applications, specific care should be taken to be sure that specular surface reflections do not degrade the available contrast. This is
easily accomplished by slightly tilting the sensor so that it is not
aimed perpendicular to the specular surface.

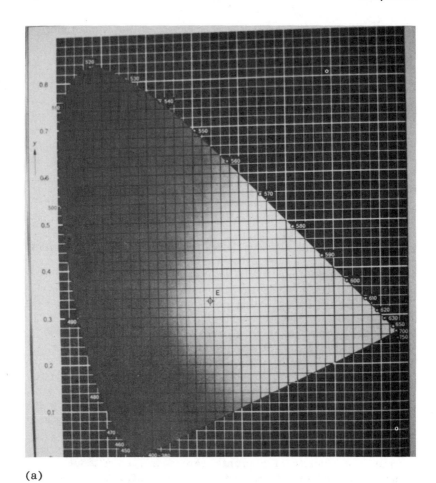

(a)

Figure 2.49 Photographs of the rainbow-colored CIE chromaticity diagram made with (a) red and (b) green light.

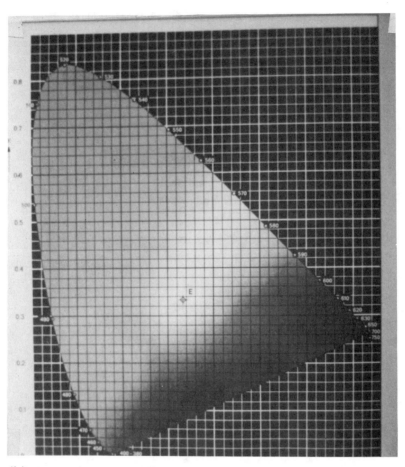

(b)

Table 2.5 Contrast Ratio of Some Common Printed Marks Measured
with Red, Green, and Infrared Light

	Contrast ratio		
Mark color	IR (880 nm)	Red (660 nm)	Green (565 nm)
Red, bright	1.02	1.08	9.5
Red, dark	1.13	1.34	11.5
Orange	1.01	1.01	3.3
Yellow	1.01	1.00	1.00
Yellow-green	1.01	4.0	2.6
Green, light	1.02	4.3	1.4
Green	1.10	7.1	3.5
Green, dark	1.24	8.5	6.1
Blue, light	1.04	4.1	2.4
Blue	1.22	10	12
Blue, navy	2.1	11	20
Purple, grape	1.05	5.2	7.7
Butterscotch	1.00	1.03	1.7
Brown, light	1.45	1.65	3.0
Brown	2.0	2.5	5.3
Brown, dark	4.7	8.2	20
Black	22	32	27

3

Photoelectric Optical Configurations

In a moment, in the twinkling of an eye.

I Corinthians 15:52

In this chapter we describe in detail the optical configurations commonly used in photoelectrics, their key properties, how these properties can be taken advantage of, and what pitfalls may be encountered in their application. Repeated success in the design or troubleshooting of photoelectric applications can only be the result of thorough understanding.

3.1 DIFFUSE PROXIMITY SENSING

Diffuse proximity sensing is the detection of the sensor's own transmitted light as reflected from the target object's diffusely reflecting surface as shown in Fig. 3.1. These sensors are intended for direct detection of the object rather than indirect detection via beam blocking. They can therefore be used without need for a reflector or detector on the other side of the target object. Both source and detector are on the same side of the object, mounted side by side, usually in the same photoelectric housing. Because direct object sensing is needed at a variety of ranges, slightly different optical designs are required to optimize the optical performance for each of these sensing ranges. Diffuse proximity sensors fall into one of four basic classifications: long range, short range, variable range, or fixed-focus proximity. The advantages of proximity sensors include: wiring

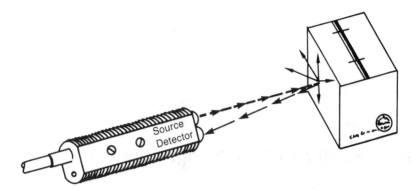

Figure 3.1 Diffuse proximity sensing is the detection diffusely reflected light from an object surface.

only one side of the sensing zone; simple alignment; and differences in surface color can be detected. The disadvantages include: limited sensing range, and range sensitivity to surface color.

3.1.1 Long-Range Proximity Sensors

Long-range proximity sensors are optically optimized for the longest possible detection range by aiming the source light and the detector field of view at a common distant point. This causes the light beam and the detector field of view to be fully overlapped at long distances, as shown in Fig. 3.2, thereby returning the maximum possible signal. Figure 3.2 also shows how the effective beam diameter (the overlap area) grows in size proportional to range. Near the lens, the effective beam is more oblong. For merged lens sensors, the effective beam becomes just a narrow line that runs the length of the seam between the source and detector lenses at near zero range. Long-range proximity sensors have a predictable falloff in signal strength that is inversely proportional to the square of the distance. This corresponds to what is called a slope of -2 on a log-log scale excess gain graph. The excess gain graph of typical long-range proximity sensors shown in Fig. 3.3a shows how they all have excess gain decreasing at the same rate as the line marked "-2 slope." Accordingly, the curve showing excess gain of 1.0 at 100 in.

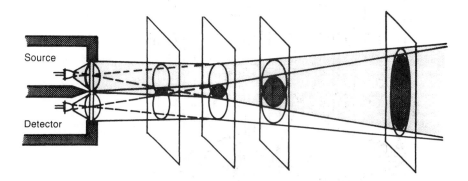

Figure 3.2 Long-range proximity source beam and detector fields of view are fully overlapped at long ranges.

would be expected to have an excess gain of 4.0 at 50 in. (one-half the range is four times the excess gain), and an excess gain of 16 at a range of 25 in. The inverse-square relationship can easily become a disadvantage. We only want to sense objects where we want to sense them, not elsewhere! Having an effective beam that extends out indefinitely, gradually growing weaker, may be satisfactory when detecting a truck in a shipping dock, but will not work on a narrow conveyor where people may pass closely on the other side. Sensing range can be controlled, to some degree, if necessary, by reducing the sensor's excess gain with the sensitivity adjustment. However, range is more effectively controlled with cross-eyed short-range sensors discussed in Section 3.1.2. The difference in excess gain between the curves in Fig. 3.3a at close range is a function of the source and detector lens separation and the field patterns of the lenses. If the source and detector lenses are separated from each other, there will be little or no overlap of the source beam and detector field of view close to the sensor, resulting in little or no excess gain at short ranges. The curve showing the lump at just under 1 in. of range is due to an artifact in the radiation pattern of the source and detector lenses probably due to multipath reflections in the optical cavity interior.

Figure 3.3b shows the sensing zone for the 100-in. long-range proximity sensor. This curve specifies at what point a large white target entering the beam from the side will be detected by the sensor. Although the effective beam diverges with the dashed

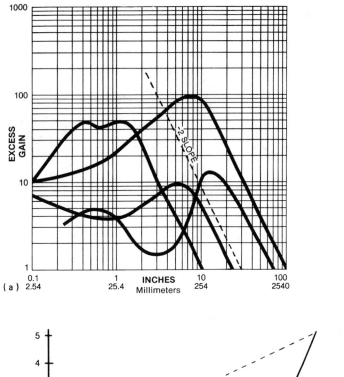

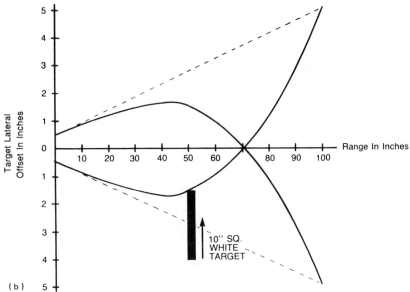

Figure 3.3 (a) Typical published excess gain curves for long-range proximity sensors; (b) effective beam diameter (dashed line) and target sensing zone diameter (solid line) for the 100-in. sensor.

line, when G_X is high, only the edge of the effective beam need
be entered to return sufficient light for detection. Similarly, we
would expect target detection to require full beam penetration at
maximum range, 50% penetration at the range where $G_X = 2$, and
25% penetration at the range where $G_X = 4$.

3.1.2 Short-Range Proximity Sensors

Short-range proximity sensors are produced in two varieties.
One way to reduce range is to use smaller optical elements. This
is often done by using lensed photoelements and replacing the
large objective lens with a transparent window. These sensors
might better be thought of as weak long-range proximity sensors since they will exhibit the same -2 slope on the log-log excess gain graphs. Alternatively, short-range proximity sensors
can be designed to have very sharp excess gain cutoff with distance in order to detect intended targets well but to avoid erron-
eous detection of background objects. There are two useful techniques for producing short-range proximity sensors. First we
examine the standard construction method followed by the more
sophisticated background suppression design.

The standard construction short-range proximity causes the
field of view of the detector and the source beam to cross over
each other close to the sensor as shown in Fig. 3.4. Since detection can occur only where they overlap to form the effective
beam, there can be no detection past the crossover point. After
studying Fig. 3.4, you would be correct to expect an excess gain
peak near the point of maximum beam overlap. Short-range proximity sensors are usually identical to their long-range brothers,
except that the photoelement separation is altered to achieve the
desired cross-eyed effect. Short-range proximity sensors have
less excess gain at long range because there is less beam overlap. They also have increased excess gain at close range because beam overlap is much better. Comparing the typical published short-range proximity curves of Fig. 3.5a to the long-
range proximity curves of Fig. 3.3a, it is clear that close-range
performance is improved and much sharper cutoff of excess gain
with distance is achieved, particularly when viewed in the linear
coordinates of the real world versus the logarithmic scale of these
graphs. Some sensors are able to achieve a cutoff slope of -5 or
better on a log-log excess gain graph. This translates to going
from an excess gain of 100 (almost anything can be seen with an
excess gain of 100) at 4.0 in., to unity excess gain at 10 in. versus an expected 40 in. for the long-range design. The sensing

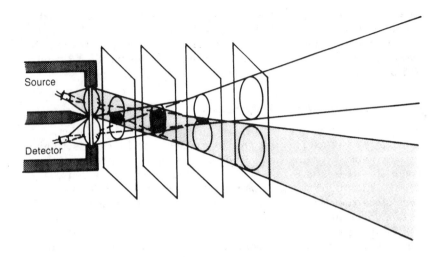

Figure 3.4 Standard short-range proximity sensors have a cross-eyed source beam and detector fields of view.

zone specification in Fig. 3.5b for the 6-in. range proximity sensor demonstrates that very small or dark targets are easily detected at 3 in. while large white targets are ignored beyond 6 in. The ability of cross-eyed short-range proximity sensors to ignore background objects has made them quite popular with photoelectric users.

The construction of a background suppression proximity sensor includes a second detector, as shown in Fig. 3.6. The second detector is electrically connected in reverse polarity and aimed so that its field of view crosses the source beam at a greater distance than that of the first detector. The range at which the source beam overlap for the two detectors is equal is where the negative detector takes over and completely suppresses detection. Beyond this range, even the best reflectors and retroreflectors will not be detected. Figure 3.7a shows the near-ideal sharp cutoff of excess gain with range. The sensing zone specification in Fig. 3.7b shows a nearly rectangular shape versus the fishtail-like shape of other proximity sensors because of the extremely sharp cutoff characteristics. When applying these sensors, the direction from which an object enters the beam may be important. To obtain the expected performance, the target object

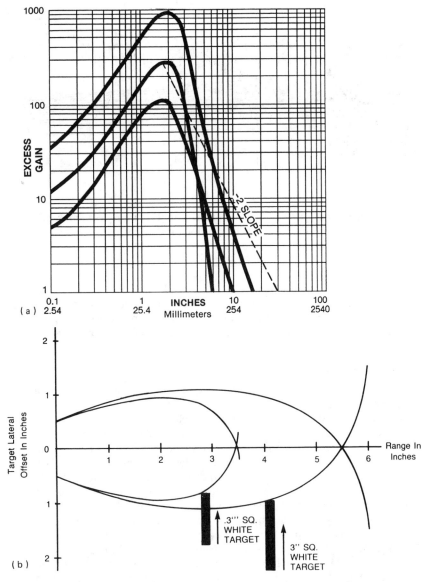

Figure 3.5 (a) Typical published excess gain curves for cross-eyed short-range proximity sensors showing steep signal cutoff; (b) target sensing zone for the 6-in. short-range proximity sensor for 3-in.-square and 0.3-in.-square targets.

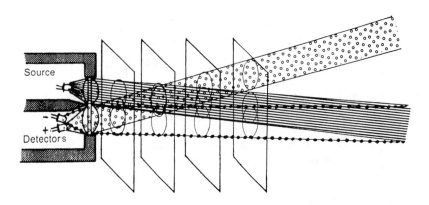

Figure 3.6 Background suppression short-range proximity sensors have a second detector to cancel totally reflections from
longer-range targets.

must simultaneously and uniformly cross into both detectors' fields
of view. This requires that the object move in a direction parallel to the imaginary line dividing the source and detector lenses.

3.1.3 Fixed-Focus Proximity Sensors

Fixed-focus proximity sensors are generally constructed with an
extended lens, as depicted in Fig. 2.3, to cause the light emerging from the lens to be focused very close in front of the sensor.
These fixed-focus sensors are sometimes referred to as convergent beam sensors. Figure 3.8 shows how the LED beam and detector field of view are focused and aimed by the lenses to form
small overlapping images of the photoelements in front of the lens.
The lens system is designed to prevent beam overlap in front of
or behind the point of beam convergence. This feature allows the
detection of very small objects or the discrimination of object
range. Some of these sensors are capable of detecting such
things as single black threads, inversion of bottle caps, and
light bulb filaments. Figure 3.9a shows the performance of some
typical focused proximity sensors. There are numerous designs
available with differing focus distances, ranges of focus, excess
gains, and effective beam sizes. The specified sensing zone in
Figure 3.9b shows that the sensing zone can be quite small and

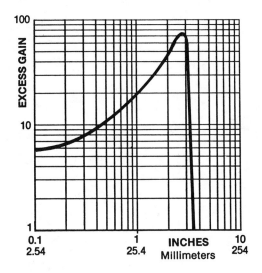

(a)

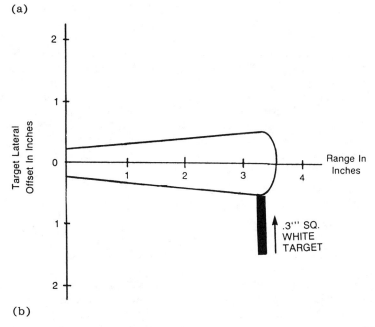

(b)

Figure 3.7 (a) Excess gain performance and (b) target sensing zone of a typical background suppression sensor.

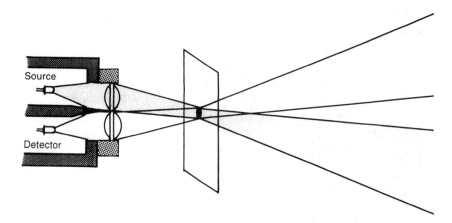

Figure 3.8 Fixed-focus proximity sensors have both cross-eyed source and detector beams as well as focused images of the photo-elements at the range of maximum beam overlap.

extends only over a small sensing range. When sensing small objects, a small effective beam size is imperative in order to be sure that the majority of signal falls on the object, increasing the light reflected back to the detector. For example, a 10% reflective black thread may intercept only 1% of a short-range proximity sensor's effective beam. To detect the thread, the sensor would require an excess gain of 1000. However, the thread may intercept 20% of a focused proximity sensor's effective beam and require only an excess gain of 50 for detection.

3.1.4 Variable-Range Proximity Sensors

Variable-range proximity sensors offer the ability to adjust the sensing range mechanically. This is done by adjusting the pointing direction of the source beam and detector field of view to control the beam overlap function. Figure 3.10 gives an idea of how such sensors can be mechanically adjusted. The relatively large separation between the device's source and detector lenses are required to produce reasonable range discrimination past a few inches of range. Figure 3.11 shows how this separation prevents beam overlap near the lenses and allows a much steeper dropoff in signal at larger ranges. On a log-log excess gain graph, the gain cutoff slope can easily exceed -5. The principal

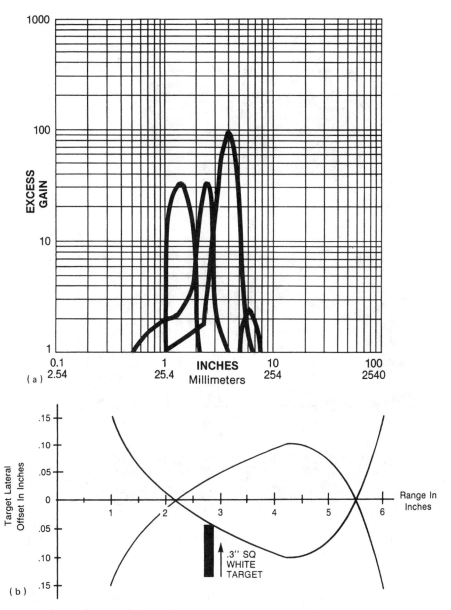

Figure 3.9 (a) Typical published excess gain curves for fixed-focus proximity sensors showing the sharp range definition; (b) target sensing zone for the 4-in. sensor showing small beam diameter.

Figure 3.10 Typical variable-range proximity sensor. (Courtesy of Opcon, Inc.)

advantage of this arrangement is that the sensor has the capability of detecting objects at relatively long ranges and ignoring even specular reflectors at just slightly longer ranges. In contrast, standard long-range proximity sensors have a gentle −2

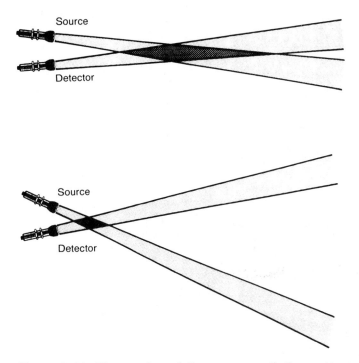

Figure 3.11 The overlap of the source radiation pattern and detector field of view for a variable proximity sensor can be set to almost any range by adjusting the pointing direction of the sensors.

slope on the log-log excess gain graph that is very prone to detecting objects passing in the near background. The excess gain graphs of Fig. 3.12 show how the sensing system pictured in Fig. 3.10 can be adjusted to sense with an excess gain of 20 at 5 ft of range and yet be quite blind past 7 ft. A long-range proximity sensor with similar excess gain at 5 ft would detect a white paper target past 20 ft. The steep-sloped excess gain cutoff at this range is entirely due to the large distance between the source and detector lenses that causes the beams to cross through each other at a significant angle rather than running parallel to each other.

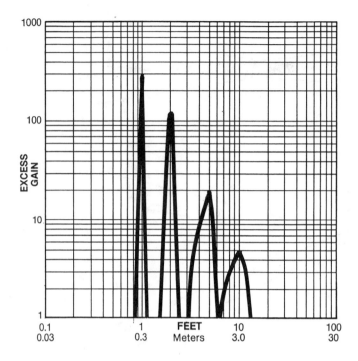

Figure 3.12 Typical excess gain curves for available high-performance variable proximity sensors.

3.1.5 Ultraviolet Fluorescent Proximity Sensors

Ultraviolet (UV) light is not visible to the human eye. UV is light that is beyond blue and violet in the spectrum and is often referred to as "black light." Ultraviolet light causes some materials to fluoresce or give off light at a different wavelength than the incident UV radiation. There are commonly available marking inks that are invisible in normal lighting but glow brightly when illuminated by a UV light source. There are also a few common materials such as cotton that fluoresce in UV light. UV fluorescent proximity sensors provide the capability of detecting the presence of these fluorescent dyes and materials that emit light in the visible spectrum when stimulated by UV radiation. Figure 3.13 shows diagrammatically how a UV fluorescent photoelectric sensor works. The UV light source is pulsed, as is the LED of a conventional photoelectric. The returned light

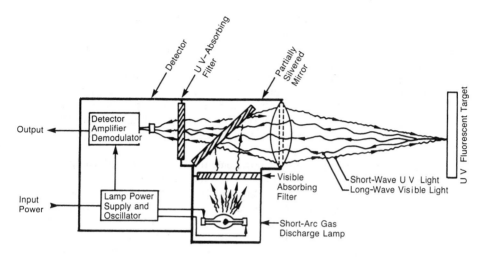

Figure 3.13 Functional diagram of an ultraviolet fluorescent sensor.

is filtered to remove any UV light. Any pulsed visible light must be the result of illuminating a UV fluorescent material. These sensors generally need only detect the general object presence and as such, use simple long-range proximity optics. Except for the special considerations for dealing with ultraviolet light, this sensor has the same optical characteristics and function as that of a long-range proximity sensor. UV fluorescent sensing is most useful when it is not practical for aesthetic reasons to mark an object visibly, or when other means of obtaining contrast for detection are not feasible. Adhesive application verification and reject marking are two typical and appropriate applications.

3.1.6 Registration Mark Detection

The discussion on registration mark detection will center on the core set of problems that are found most commonly with this detection problem but certainly apply as well to related low-contrast detection problems. Registration marks are also referred to as index marks and color marks. Registration marks are the little colored stripes found at the end of bread bags, salad dressing mix packages, toothpaste tubes, and similar packaging containers. The purpose of a registration mark is to provide a synchronizing signal to the packaging machine. The mark is detected

by a sensor that signals the machine to cut the material or seal
the package. Without this signal, the pretty picture or company
name would soon drift into the cutting area, leaving a package
that would be hard to identify and would not impress prospective
buyers. One of the greatest challenges for an optical sensor is
to distinguish this mark reliably from the surrounding package.
Product manufacturers do not want to print an extra color of ink
on the package just for registration sensing. They prefer that
the sensor be capable of detecting any color of mark on any color
background.

There is a unique set of requirements for the optical design
of a registration mark sensor. The sensor must have a small effective sensing area relative to the size of the mark in order to
obtain optimum contrast (see Fig. 2.46). Unlike the focused proximity, it must not be sensitive to slight changes in sensing range
that could be caused by flutter in the web of packaging material.
It must also be able to detect virtually any color change. Figure
3.14 shows one method by which these problems have been addressed. Switching from a red to a green LED source is provided to allow better contrast with some colors, as indicated in
Table 2.5. Changing colors is simply a matter of flipping a switch.
The partially silvered mirror causes the appearance of both LEDs
to be in the same optical position. The photodetector is aligned
coaxially with the LEDs using a second partially silvered mirror.
The same objective lens is used to focus both the source beam
and detector field of view onto the same target. The use of coaxial beams produces consistent beam overlap and reduces sensitivity to range variation. Earlier designs used primarily unmodulated incandescent illumination and ac-coupled amplifiers.
These older designs are still spoken of highly by users and seem
to have better sensitivity with the most difficult colors, such as
yellow on white. To detect any color or contrast change with
solid-state LED-based sensors will require designs that provide
even more color illumination choices than are presently available.

There have been other approaches to solving this problem.
Registration mark detection has also been successful with thru-
beam sensors on most transparent or translucent materials with
opaque registration marks. In some circumstances a standard
thru-beam sensor may be used when the contrast ratio is sufficiently high. Fiber optic sensors have also been widely used for
registration mark detection. Fiber optic cables have the advantage of being able to fit into small places where standard sensors
might not fit. By their very nature, fiber optic cables also provide small sensing apertures. However, neither thru-beam nor

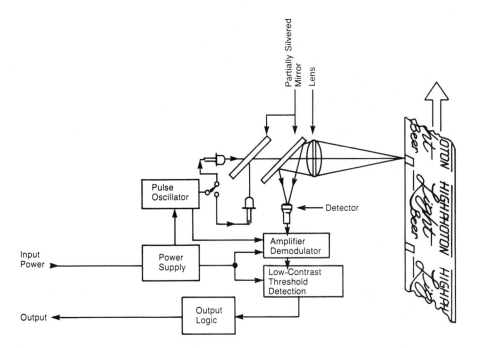

Figure 3.14 Functional diagram of a multicolored LED registration mark sensor with coaxial optics.

fiber optic sensors are designed to change the color of illumination readily and should be considered only for those applications where the object colors remain consistent. Low-contrast detection circuitry is available for some fiber optic and thru-beam sensors and should be used for improved system stability and reliability.

3.2 REFLEX SENSING

Reflex sensing is the detection of retroreflected light from a corner cube or beaded reflector as shown in Figure 3.15. These sensors are used with a reflector mounted on the far side of a passing object. Light sent out from the sensor is returned to the sensor by the retroreflector. They are intended for indirect detection of an object via beam blocking rather than direct reflective detection. These sensors are also referred to as retroreflective

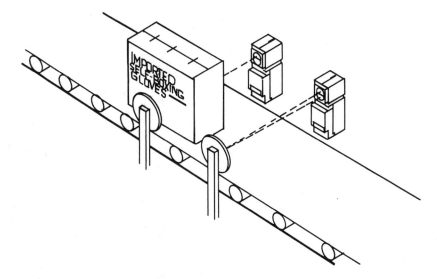

Figure 3.15 Reflex sensing is the indirect detection of objects via interruption of a retroreflected beam. Reflex sensors should not respond to diffuse reflecting objects.

sensors. The advantages of a reflex system include: no problem with inadvertent detection of objects beyond the intended sensing zone; long sensing range; no sensitivity to the object range or color changes; high light/dark contrast ratio; and easy installation and alignment. A well-designed reflex sensor should never exhibit proximity detection from any diffusely reflecting object.

3.2.1 Standard Reflex Sensors

The standard reflex sensor uses a highly efficient infrared LED as its light source. All reflex sensors have effective beams much like that of the long-range proximity sensor in Fig. 3.2. The difference is that the reflex sensors do not use immersion-lensed photoelements as do their long-range proximity brothers. This results in a smaller effective beam area due to less source and detector field of view overlap, and reduces the excess gain to the point where it will no longer respond to a diffuse reflecting surface. However, since retroreflectors return over 5000 times as

much light as a diffuse white reflector, retroreflector targets are
still easily detected.

Figure 3.16a shows the excess gain curves for a reflex sensor
when used with various retroreflective targets. Plastic corner
cube retroreflectors all exhibit the same excess gain curve function close to the sensor, but depart from one another at a few
feet of range. The range at which they start their downward
departure from the highest curve is the range at which the effective beam becomes larger than the reflector. An infinitely
large reflector or large array of reflectors constructed as shown
in Fig. 3.17 will exhibit a dropoff in excess gain with distance
that is inversely proportional to the square of the distance, much
like a long-range proximity sensor. This produces the characteristic -2 slope gain dropoff on a log-log excess gain graph.
However, most reflex sensors use a fixed-size retroreflector target. As the effective beam diameter grows larger than the reflector, the proportionate amount of signal that can be returned
is further reduced by another $1/(range)^2$ factor. The net result is that the excess gain will decrease as $1/(range)^4$ after the
effective beam diameter grows larger than the target. This produces the characteristic -4 slope tail on reflex sensor excess gain
curves. The same applies to the beaded retroreflectors. Typical
beaded retroflector return 200 to 1000 times more light than
diffuse white reflectors. The beaded reflector target graphed in
Fig. 3.16 is rated at a reflectance of 200. At long ranges, Fig.
3.16 confirms that its reflectance is only about 1/40 that of a
plastic corner cube retroreflector. The explanation for the difference in the shape of the excess gain curves at close range is
based on two differences in the reflectors. Most notable is the
physical size of the corner cube cell relative to single glass beads.
Figures 2.12 and 2.13 imply that the returning beam, although
parallel, has been translated slightly sideways. The slight sideways translation of the beam results in a virtual increase in the
effective beam width and provides tremendously increased coupling at close range. The small size of the glass beads does not
produce significant lateral beam translation and does not result
in increased coupling. The second difference is that the observation angle of a corner cube reflector is much narrower than
that of a beaded reflector. Figure 2.15 indicates that the half-
power observation angle for a beaded reflector is about five times
as wide as that of a corner cube reflector. The reflector range
and the distance between the lens centers determines the observation angle. An ideal retroreflector returns all the light energy
to its original source. A photoelectric sensor requires that some

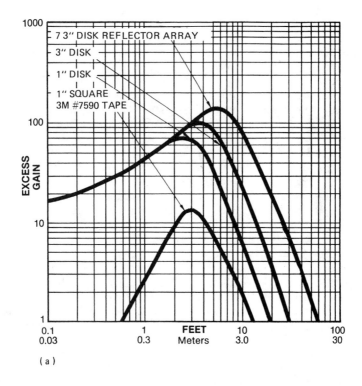

Figure 3.16 (a) Typical excess gain curves of a reflex sensor for a variety of retroreflective targets; (b) their related target sensing zones.

of this energy return to the adjacent lens for detection. A beaded reflector will return light that will fully encompass the detector lens at about 1 ft of range while the corner cube reflector must be over 5 ft away before its returned light fully includes the adjacent detector lens. These two factors, along with the beam overlap function, are responsible for the determination of the close range coupling characteristics and the range at which excess gain peaks.

The target sensing zone for reflex sensors in Fig. 3.16b shows how much of the effective beam must be blocked by the target before the retroreflected light is reduced below detection threshold. The effective beam of a reflex sensor is defined not only by the overlap of the LED radiation pattern and detector field of view,

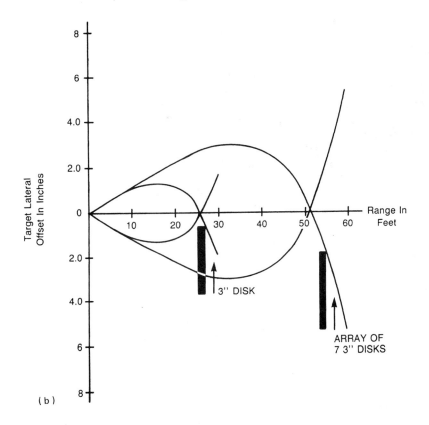

(b)

but also by the size of the retroreflective target. In retroreflective sensing, an object can block the returned signal only if, in
fact, the signal was returning in the first place. The beam overlap sensing zone may be larger than the retroreflector allowing
potentially returned signal to spill over the sides of the reflector.
The effective beam for reflex photoelectrics is never larger than
the sensor lens area at close range, never larger than the retroreflector target boundaries, and never larger than the beam overlap area at the target. Figure 3.18 shows how the lens and retroreflector constrain the size of the effective beam. When an object must reliably break the beam over a very short distance, the
shape of the retroreflective surface can be altered to reduce the
width of the effective beam. This can be accomplished by mask-
ing a retroreflector or trimming it to the required dimensions.

Array of 19

Array of 7

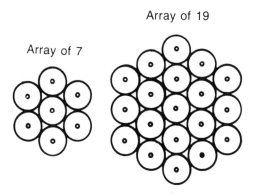

Figure 3.17 Large retroreflective targets made with multiple retroreflectors can increase the range capability of reflex sensors manyfold.

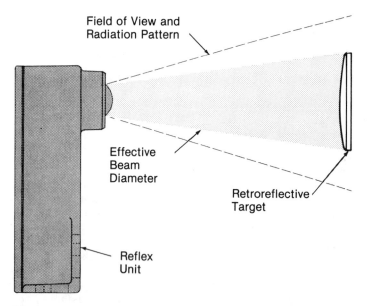

Field of View and
Radiation Pattern

Effective
Beam
Diameter

Retroreflective
Target

Reflex
Unit

Figure 3.18 Limitation of a reflex sensor's effective beam by the lens size and reflector target size. (Courtesy of Opcon, Inc.)

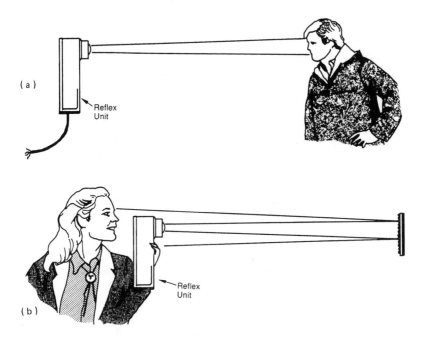

Figure 3.19 The visible beam location may be determined by (a) looking back at the sensor and finding the position of brightest lens glow, or (b) by observing the glow on the retroreflector when viewed from directly behind the sensor.

Most often there is sufficient excess gain in the system, so that if only 10% of the available light was returned, operation would still be reliable in reasonably clean environments.

3.2.2 Visible Beam Sensors

Visible beam reflex sensors operate on the same principles as infrared sensors. Their principal advantage over the infrared variety is the ability to see where the beam is so that it may be more easily aligned with the retroreflector. All present visible beam reflex sensors are manufactured with red LEDs. Other visible LEDs do not produce as much light energy as do red LEDs. Except at very close ranges, the LED beam cannot be seen on a white paper target in standard room lighting. However, Fig. 3.19 shows two useful techniques to make use of the visible beam for

alignment. The light from the sensor can easily be seen by the
unaided eye over 100 ft away when looking on axis toward the
sensor. Looking back at the sensor into the beam is quite useful for visually determining where the beam is pointed. When
your eye is right on the axis of radiation, the full area of the
lens glows bright red. The second method takes advantage of
the highly reflective nature of retroreflectors. If you put your
eye behind the sensor so that you are looking down almost exactly the same path as the LED light beam, you will see the retroreflector glow distinctly red when the beam crosses over the reflector. If your eye is not almost directly behind the sensor,
your observation angle will be too large and you will not see the
retroreflection. Some visible beam sensors are not bright enough
to be visible by this method when the retroreflector is more than
a few feet away.

There are two possible liabilities that visible beam reflex sensors carry. The best visible LEDs have only about 10% of the
signal strength of a good infrared LED. This means that the
visible beam sensors will not have the range or excess gain that
could be expected from their infrared counterparts. This is important in mild to dirty environments. Second, visible beams
can be seen. This may be a problem when a visible beam might
attract vandalism.

3.2.3 Polarized Reflex Sensors

Under certain conditions, polarization can be used to advantage
in the application of photoelectric sensors. A typical problem in
the application of reflex photoelectric sensors is that objects
blocking the retroreflector may have a near-mirror-quality specular finish. A specular perpendicular surface without about 1°
of the sensor's optical axis can often fool the sensor with retro-
reflector-like properties. By using the properties of polarizing
filters and knowing that a specular reflection does not alter the
plane of polarization, we are able to cause specularly reflected
light to be absorbed by inserting cross-polarizing filters in front
of the sensor's photoelements as shown in Fig. 3.20. Although
a corner cube retroreflector also uses the principle of specular
reflection, it will return light successfully through the cross-
polarized filters of the sensor. The plastic injection-molded corner cube retroreflector is filled with a lot of residual stress from
the differential cooling rates between the sharp corner surfaces
and the interior. The stressed plastic becomes birefringent and
rotates the polarization plane of incident polarized light, enabling

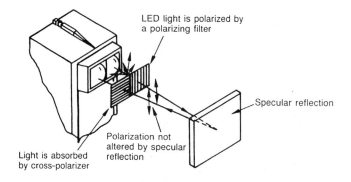

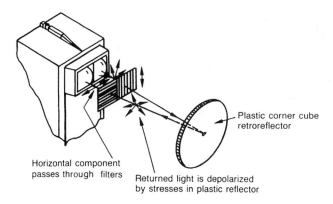

Figure 3.20 Elimination of first surface reflections using polarized filters.

it to pass through the second cross-polarized filter to the sensor detector. Glass-beaded retroreflectors do not alter the polarization of incident light and will not function with polarized sensors.

Figure 3.21 shows the excess gain response of a number of common specular reflecting objects using the optics of a typical reflex sensor. Each of these items was carefully aligned to achieve the greatest possible returned signal at each range. Each of these items reflects sufficiently at some range when properly aligned to trigger the sensor. An observation worth pointing out is the difference in reflection characteristics between the mirror and soft-drink can, and the glass plate and glass jar.

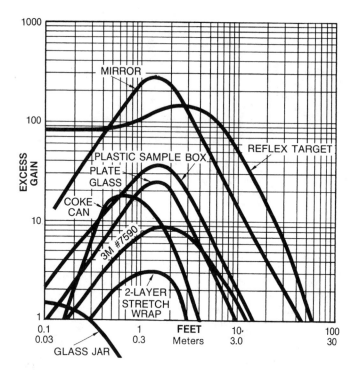

Figure 3.21 Signal strength returned from various targets as measured with typical unpolarized reflex sensor optics.

Curved or otherwise distorted specular surfaces produce much less signal than do their flat-surfaced counterparts. The cross-polarized filters reduce these first surface specular reflection magnitudes sufficiently to prevent inadvertent detection at any orientation. Presumably the result is carefree application and installation. However, polarization is not a magic cure for the speculars. Polarized sensors work well when understood and applied properly. Polarized sensors have clear limitations as well. Excess gain is cut by about a factor of 10, and range by about a factor of 2 when polarizers are installed. Polarized visible beam reflex sensors have such low excess gain that they would probably not detect any of these reflections, except the mirror, strictly on the grounds of low excess gain. Reduced excess gain may present a problem in moderately dirty environments.

Figure 3.22 Stretch wrap and other plastics rotate the polarization of light passing through it. Polarized sensors provide no
advantage when used with these materials.

Infrared polarizers are available but do not provide the excellent
cross-polarization attenuation available from visible polarizers.
Infrared polarized sensors generally ignore all specular surfaces
except a mirror, unless the sensitivity is reduced a bit. The
worst popular misconception is that cross-polarized sensors work
with plastic materials. Stretch wrap is an excellent depolarizer.
Figure 3.22 shows a piece of stretch wrap positioned between
cross-polarized filters. Figure 3.23 shows the reflection of a

(a)

(b)

Figure 3.23 (a) Reflection of a lamp on a mirrored surface
with stretch wrap pulled across it. (b) When the lamp illumination is polarized and a cross polarizer placed in front of
the camera, only the light passing through the stretch wrap
is received.

lamp on a mirror surface with stretch wrap pulled across it. Figure 3.23b used cross-polarized filters to eliminate the reflection of the bulb. However, light passing through the stretch wrap had no problem passing through the second polarizing filter to the camera. Clear plastic sample boxes tested also easily fooled the best polarized reflex sensors; often without a backing mirror surface for help. Under the right conditions it is even possible to false trigger a polarized reflex sensor with a single piece of transparent injection-molded plastic because the specularly reflected light from the back surface of the plastic has traveled through the plastic. A polarized sensor is not a solution when transparent plastic is involved! The key solution provided by a polarized reflex sensor is elimination of first surface reflections, but not reflections from any surface below the first. With the exception of the mirror, each of these reflections can generally be avoided by reducing the sensitivity of the gain adjustment. All of them can be avoided without polarization by tilting the sensor about 10° away from perpendicular to the surface.

3.3 THRU-BEAM SENSING

Thru-beam sensing is the detection of light transmitted directly from a source to a detector as shown in Fig. 3.24. Thru-beam sensors utilize indirect detection of an object via beam blocking rather than direct or indirect reflective detection. These sensors are also referred to as opposed, break beam, transmitted beam, and thru-scan; the LED source as the emitter, sender, or transmitter; and the detector as a receiver. Thru-beam sensing advantages include: no problem with inadvertent detection of objects beyond the intended sensing zone; extremely high excess gain, making it the best choice for most dirty applications; longest sensing range; greatest light/dark contrast ratio; smallest possible effective beam; and not sensitive to the object distance or color changes. Disadvantages include: two components to wire across the detection zone; greater system expense; and an often difficult alignment process.

3.3.1 Standard Thru-Beam Sensors

Thru-beam sensors require only one optical element on each side of the detection zone. However, thru-beam sensors are commonly packaged in the same housings as are reflex and proximity sensors and use the same double objective lens. The majority of

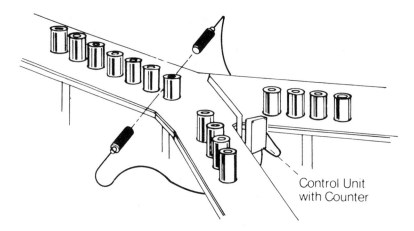

Control Unit
with Counter

Figure 3.24 Thru-beam sensing is the indirect detection of objects via interruption of a transmitted beam of light. (Courtesy of Opcon, Inc.)

thru-beam sensors today use an infrared LED source. However, the trend is toward including product models with a visible red beam in spite of the inherent poorer optical performance, because the visible beam aids in alignment. Some of the source sensor packages with the double objective lens take advantage of the second lens by using a visible LED behind one lens and an infrared LED behind the other. This provides the double advantages of alignment ease and optical strength.

The effective beam and excess gain relationships for thru-beam sensors are the easiest to understand of all optical sensor varieties. The effective beam in a thru-beam system is not as large as the field of view of each optical beam. Rather, it is like a rod that extends from the source lens to the detector lens, as shown in Fig. 3.25. Similarly, when you look toward a light bulb, there are many places your thumb may be placed so that the light bulb will illuminate it and simultaneously be in the field of view of your eye. However, only when your thumb is directly between the bulb and your eye will the bulb light be interrupted. Object detection occurs in a thru-beam system when the source light is interrupted. How much light must be blocked is a function of the net operating margin of the system. If the system has sensors positioned to give them excess gain of 400 and there is 50% dust coverage on each lens, the net

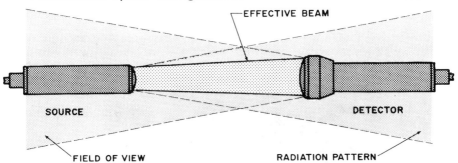

Figure 3.25 The effective beam of a thru-beam sensor is like a rod that extends from the source lens to the detector lens. (Courtesy of Opcon, Inc.)

operating margin is 400 × 0.5 × 0.5 = 100. This means that the beam will not be blocked sufficiently until it is more than 99% obstructed. If the starting excess gain were only 8.0, the net operating margin would be only 2.0. At a net operating margin of only 2.0, sufficient beam blocking is obtained with slightly less than 50% beam obstruction. In the second case, the trip-point position for the object blocking the beam would be highly sensitive to the amount of dust on the lens. This is one reason why apertures are often useful on thru-beam sensors. An aperture provides control over the shape of the effective beam. The slit aperture shown in side view in Fig. 3.26 creates a ribbon-shaped beam. Objects crossing the beam via the narrow dimension will have highly repeatable sensing positions. The value of the longer dimension is that it provides more lens surface area for transmission and reception of light resulting in greater excess gain. Small objects may also require an aperture over the sensor lenses in order to reduce the effective beam dimension so that they will be capable of complete beam blockage.

Another important function of apertures is the control of multipath detection of the source beam. Figure 3.27 shows that it is possible for alternate transmission paths of the light to reach the detector. These alternate transmission paths are not always repeatable and lead to intermittent malfunction. Alternate paths can come from surrounding pieces of the machinery, support structure, or an adjacent object close behind the object presently breaking the beam. An aperture will restrict the field of view of both the source and detector so that the alternate path is eliminated.

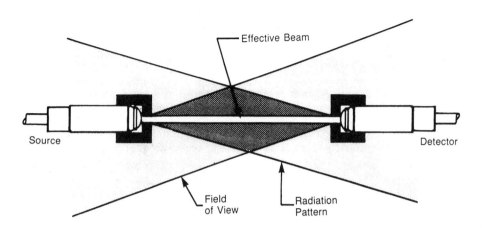

Figure 3.26 Apertures are used to change the shape of the effective beam for improved detection characteristics.

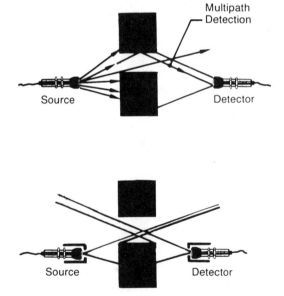

Figure 3.27 Apertures are used to reduce the field of view of the source and detector to prevent multipath detection.

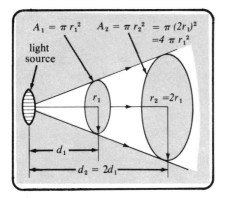

Figure 3.28 Signal received by the detector is inversely proportional to the square of the distance between the source and detector. (Courtesy of Opcon, Inc.)

The excess gain of a thru-beam system is the simplest of all the sensor types to understand. Figure 3.28 diagrammatically shows that by doubling the range from the source, the light energy produces a spot with twice the diameter. The ratio of the areas is the ratio of the square of the diameter. In this case that would be four times the area. Since the amount of light energy remains constant but is spread over four times the area, a detector lens with fixed area would receive only one-fourth the energy at twice the distance from the source. At three times the range, only one-ninth of the signal will be received. The signal received is inversely proportional to the square of the distance. In fact, given the maximum sensing range of the thru-beam system, the excess gain at any range can be calculated easily via Eq. 3.1:

$$\text{Thru-beam } G_X = \frac{(\text{maximum range})^2}{(\text{range})^2} \qquad (3.1)$$

When plotted on a log-log excess gain graph, it will be a line with a slope of -2. Another observation made from Fig. 3.28 is that the same signal can be received at twice the range if the lens diameter is doubled, giving four times the light collecting area. Similarly, the range is again doubled by doubling the diameter of the source lens. Given the same LED and photodetector, the excess gain will be 16 times greater or the range four times greater if lenses twice the diameter are used. The graphs

of the excess gain curves shown in Fig. 3.9a bear out this relationship for thru-beam systems having different lens diameter options. As we saw in Chapter 2, a larger lens creates a tighter
conical beam. A tighter conical beam concentrates the available
pulsed light into a smaller divergence angle, which gives it greater
sensing range at the expense of alignment ease. The smallest-
diameter lensed sensor has a beam diameter that quickly grows
larger than that of the larger-lensed units. The size of the light
beam diameter is an indication of how easy alignment will be, not
the size of the effective beam. The effective beam remains the
same size as the lens diameter at all ranges between the source
and detector. The thru-beam sensing zone specifications for the
50-ft and 200-ft sensors are shown in Fig. 3.29b. When the
range between the source and detector is small and the excess
gain quite high, almost the entire effective beam must be blocked
for object detection. At 71% of maximum range, where excess
gain is only 2, only half of the effective beam need be blocked
to attain object detection.

Thru-beam sensors are occasionally used to detect contents
within a container. Many containers, such as plastic bottles and
paper boxes, are diffuse light transmitters. Light entering this
material is scattered within and exits in a randomly directed
Lambertian distribution. When the light from a sensor falls on
one surface, a fuzzy spot appears on the opposite side. One
can estimate the spot diameter as being the sum of the diameter
of the incident light spot plus the distance between the two surfaces. For example, a 1/2-in.-thick (including the interior gap)
soap bottle with a 1/2-in.-diameter light spot projected onto it
would produce a fuzzy glowing spot about 1 in. in diameter, as
shown in Fig. 3.30. Since the emerging light is Lambertian distributed, the rules of thumb for thru-beam coupling do not apply. Other rules apply for determining the best optics for penetration of the diffuse transmitting material. To better understand what detector optics would give the best performance,
let's first consider how different detector optics respond to a
large uniformly illuminated target. Imagine your eye as the detector and the wall of a sun-lit white house as the target. Whether
you stand 5 ft from the wall or 2 ft from the wall, you still only
see a white wall of the same brightness. Your eye receives the
same amount of light in both cases even though the distance and
amount of visible wall has changed. Now, consider the case of
standing 5 ft away from the white wall and comparing its apparent surface brightness when viewing the wall with your naked
eye versus viewing it with binoculars. Similarly, the binoculars

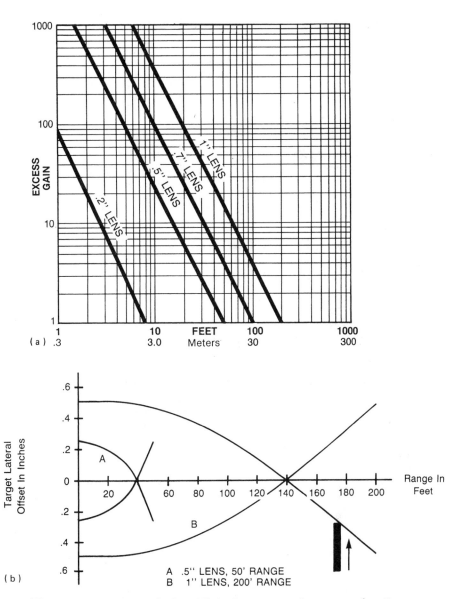

Figure 3.29 (a) Typical published excess gain curves for thru-beam sensors showing characteristic -2 slope and range proportional to the square of lens diameter; (b) sensing zone specification of thru-beam sensors.

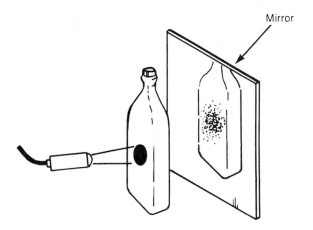

Mirror

Figure 3.30 A source beam spot on the near side of diffuse transmitting material produces a larger fuzzy spot on the far side with
a Lambertian radiation pattern.

do two things simultaneously. First, they expand a small part of
the wall image to fill our field of view (as if we were closer). Second, the large lenses collect more light from this small part of the
wall. The effects of collecting more light, but from a smaller image
area, compensate to produce an image with the same overall brightness.

Relating this back to our sensor detector, as long as the light
intensity is uniform over the field of view, large- and small-lensed
sensors (of equal f-number) will each receive the same amount of
signal. In addition, the signal strength remains constant with distance, as shown in Fig. 3.31, as long as the light spot is larger
than the detector's field of view. At ranges where the detector's
field of view becomes larger than the light spot, the detected signal strength falls off as the inverse square of the distance. Given
that you have determined the preferred mounting distance from
the target, the optimum signal coupling can be achieved by ensuring that the detector's field of view is equal to or smaller than
the light spot, regardless of the light-collecting area of the lens.

There is an important principle called "reciprocity" which applies to the swapping of source and detector positions. The principle of reciprocity states that the positions of the source and detector can be interchanged with no change in signal coupling. The
interchange requires that beam patterns are also swapped. If 1%

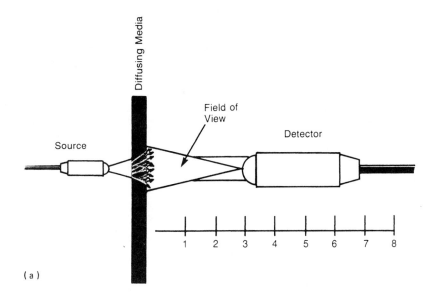

(a)

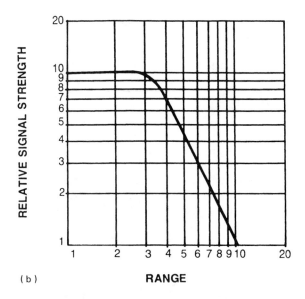

(b) **RANGE**

Figure 3.31 (a) Detector field of view relative to spot size determines (b) signal strength versus range.

of the source light is diffusely transmitted toward the detector within its field of view, it makes no difference to the detector how the light is distributed; the average intensity over the field of view will be the same. This implies that the distance of source does not matter as long as the spot of light it creates is within the detector's field of view. This is the same result as that obtained previously for the detector. As a practical matter, the detector field of view is generally a bit larger than the source radiation beam pattern because the photodiode is generally a little larger than the LED and hence will produce a slightly wider field of view. Also, the LED does not produce a uniform light spot. As a result, reciprocity cannot be applied directly, but is a close approximation to actual results.

If we were to fix the source and detector sensor heads and move a translucent (diffuse transmitting) piece of paper from the source toward the detector, we would obtain the results of Figs. 3.32 and 3.33. When the optics of both the source and detector are the same, as in Fig. 3.32, the maximum coupling occurs where the LED spot diameter and detector field of view are the same. When the optics of the source and detector are quite different, as in Fig. 3.33, maximum coupling occurs when the translucent object is near the sensor head with the widest beam pattern. Maximum coupling does not occur where the beam diameters are the same in this case because there are other minor secondary beam components that become significant very near the small sensor heads. Figure 3.33 is a pretty clear confirmation of reciprocity.

3.3.2 Edge Guide Control

An edge guide control is a minor variation on a reflex sensor that allows two reflexlike units to operate with each other as a dual thru-beam sensor. Normally, the detector of a reflex sensor will only detect pulses that are synchronous with the LED pulse generator. Here the detector circuit is operated asynchronous from the LED pulse generator. This allows the receiver to detect reliably the pulsed LED of other similar units. The receiver and optical properties of the edge guide control are like those of normal thru-beam sensors. The two beams are used to determine if a guided web has deviated outside the limits of acceptable travel. Figure 3.34 shows how the edge guide control operates. One beam is normally blocked, and the other normally complete. A change in this status causes one of the control outputs to change state and activate a device that will alter the path of the web back toward center position.

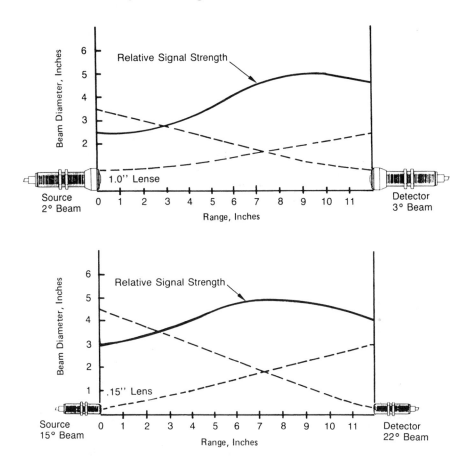

Figure 3.32 When source and detector optics are the same, the relative detected signal strength of a thru-beam system with a translucent interposed diffuser is maximum when the diffuser is located where the fields of view of the source and detector are the same diameter.

3.3.3 Data Transmission

Serial data transmission can be accomplished with thru-beam source and detector units that have special communications interfaces built into them. The pulsed LED is gated on and off by the incoming signal. The modulated LED pulse bursts are

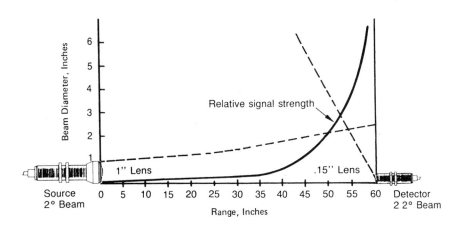

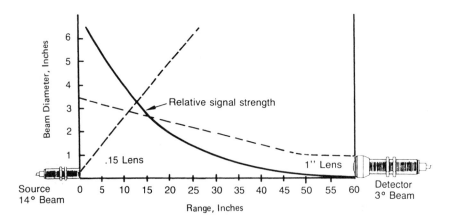

Figure 3.33 When source and detector optics are different, the relative detected signal strength of a thru-beam system with a translucent interposed diffuser is maximum when the diffuser is located nearest the end with the widest field of view (smallest lens).

demodulated by the receiver to reconstruct the original signal. These devices are used for transmission of computer data to moving stacker cranes and other fixed-track vehicles. Remote-control transmitters are also available for sending instructions to machinery such as overhead cranes. Pulse codes are transmitted

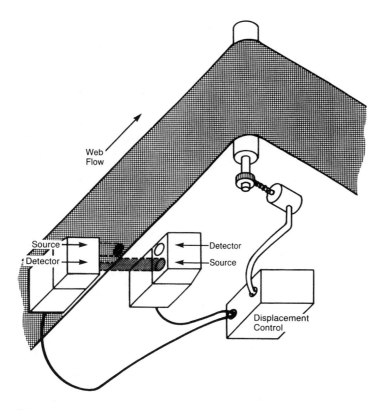

Figure 3.34 Edge guide controls have two adjacent thru-beam sensors.

by an infrared LED, following the pushing of a button to des-
ignate the action desired. Remote control transmitters and receivers work on the same principles as thru-beam sensors. They
do, however, often employ considerably different optics. Multiple emitters and detectors are typically used in order to gain the
desired receiver field of view or transmitter radiation pattern.

3.4 FIBER OPTICS

Fiber optic sensing is the detection of light transmission from a
source LED through an optical fiber cable to a reflecting target
and returned through a second optical fiber cable to a detector

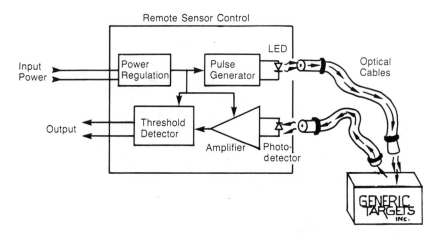

Figure 3.35 Fiber optic sensing is reflex, proximity, or thru-beam sensing with the use of intermediate light-conducting cables to convey light to and from the sensing location.

as shown in Fig. 3.35. Fiber optics are used in any of the afore-mentioned sensing modes: proximity, reflex, or thru-beam. The advantages of fiber optics include: ability to get into very tight places; very small or odd sensing shapes; high operating temperatures to 450°F and higher; immunity to electrical noise; no electronics at the sense point; and explosion-proof by their very nature. The disadvantages include: limited sensing range; increased sensor cost (although total installed cost may be less); sensitive to dirt or moisture on the fiber sensing tip; and fragility of the glass fibers.

3.4.1 Thru-Beam Sensors

Thru-beam fiber optics makes use of two separate fiber cables. One cable is attached to the LED source and the second is attached to the photodetector as shown in Fig. 3.36. The fiber cables are fairly easy to align since they typically have an exit cone and acceptance cone of a little over 30° in any direction from the fiber axis. Figure 3.37 shows the exit and entrance cones of the fibers and the effective beam between the fiber tips. As with standard thru-beam sensors, the effective beam is a rod that extends from the end tip of the source cable to the end

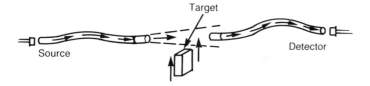

Figure 3.36 Thru-beam fiber optic sensing.

tip of the detector cable. Only when this effective beam is interrupted will the sensor detect a change. The entrance and exit
cones are analogous to the angular LED radiation pattern and the
detector field of view. The 30° exit and acceptance cones are the
maximum possible due to the limitation of the critical angle for
total internal reflection within the fiber. However, this does
not imply that there will be light exiting at this maximum angle
or that the detector will be sensitive at this angle. Figure
3.38 shows how fibers mounted very close to photoelements will
generally make use of the full exit and entrance cones. Fibers
mounted at a small distance from the photoelements will have light
enter only at a small angle. Since the angle does not grow steeper

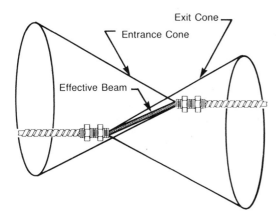

Figure 3.37 Entrance cone, exit cone, and effective beam for a
thru-beam fiber optic sensor.

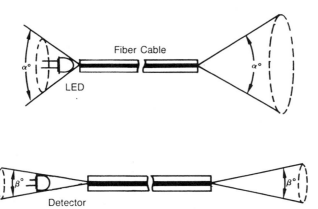

Figure 3.38 Fiber optic sensor field of view is determined by the proximity of the photoelement mounting to the fiber cable.

or more shallow during transmission, light will exit only at the same angles that it entered. The reverse is true for the detector cable. A photodetector at a distance will receive light only in a defined small conical angle. If light enters the cable at a larger angle, it will exit at the same large angle and miss the photodetector. The entrance and exit angles or field of view will vary from design to design and are best determined from the individual product specifications.

As with standard thru-beam sensors, the field of view can be controlled by the use of apertures or field stops. This can be useful to prevent multipath detection when light reflects from a second object around the blocking object and defeats the detection. Apertures for control of the sensing area are generally not used since the variety of standard fiber tip shapes is numerous. Special configurations are easily built to order by many fiber optic cable manufacturers. Also, as with standard thru-beam sensors, it is possible to increase the excess gain and range of a system by attaching lenses to the fiber tips. As a general rule, the excess gain will increase by the ratio of increased area from fiber cable to lens (assuming that the lens f-number matches the fiber entrance and exit cones). For example, a typical 0.125-in.-diameter cable with an attached 0.50-in.-diameter lens on both the detector and source cables would give an area increase of $4^2 = 16$ on both sides, for a net increase

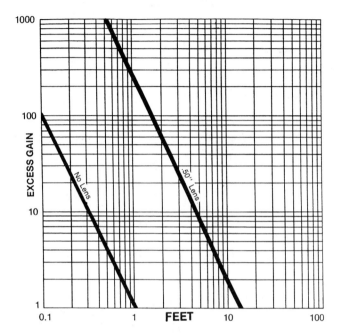

Figure 3.39 Excess gain characteristics for lensed and unlensed fiber optic thru-beam sensors.

in gain of $16^2 = 256$. Note that it takes about a 0.70-in.-diameter body to hold a lens with a useful diameter of 0.50 in. The nature of range to excess gain for thru-beam is the inverse-square relationship. This means that the range would be increased in our example by a factor of 16. Figure 3.39 shows the relative performance for lensed and unlensed thru-beam fiber tips, each having the expected -2 slope on the log-log excess gain graph.

3.4.2 Proximity Sensors

Proximity sensing with fiber optics makes use of a "Y"-branched fiber cable referred to as a bifurcated cable. Bifurcated cables combine the fibers from the source and detector together in a random mix at the sense point as shown in Figs. 2.41 and 3.40. The bifurcated cable provides the advantage of an extremely

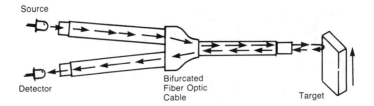

Figure 3.40 Proximity sensing with bifurcated fiber optics.

compact sensing tip. The mixed source and detector fibers in the sensing tip provide for a reasonably uniform overlap of the overall exit and acceptance cones. As a result, excellent performance and predictability of proximity sensing can be achieved down to incredibly small ranges.

To understand the excess gain performance of a proximity fiber bundle, consider first an extremely large diameter bundle. Near the center of the bundle, the light intensity of the target surface will remain constant at ranges small compared to the bundle diameter. This means that the light returning to the detector fibers will be constant over this range and would produce a horizontal excess gain curve. Next, consider a target at a distance that is large compared to the bundle diameter. Here both fields of view still overlap, but the fractional conical area of the detector bundle relative to the total cone area that the diffuse Lambertian reflected light from the target radiates into diminishes by the inverse square of the distance. The net result expected for the entire excess gain function would be a -2 slope curve at large distances that flattens to a slope of about -1 at a distance about the diameter of the cable tip and flattens further toward a slope of zero at very small distances. Figure 3.41 shows these characteristics for a typical 0.125-in.-diameter fiber tip. The excess gain curve of a flat, wide tip would spend a little more time going through the -1 slope region because of its asymmetrical shape. As expected, the sensing zone specification is very similar in nature to that of the long-range proximity sensor. Because the source and detector fibers are packed so close to one another, caution must be taken to prevent the buildup of dust or other material on the fiber tip, as it will easily couple light back to the detector and foul the detection process.

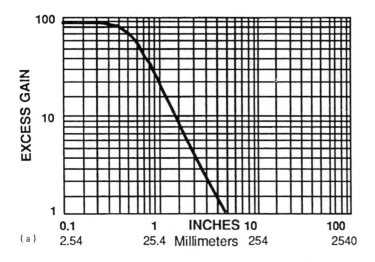

(a)

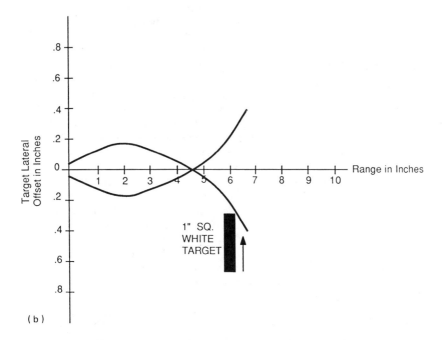

(b)

Figure 3.41 (a) Excess gain and (b) sensing zone characteristics for typical bifurcated fiber optic proximity cables.

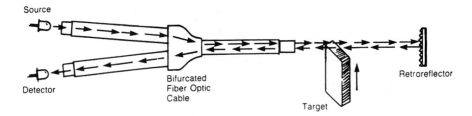

Figure 3.42 Reflex sensing with bifurcated fiber optics.

3.4.3 Reflex Sensors

Reflex operation with fiber optics is possible. It is most simply
done with the standard bifurcated proximity cable aimed at a
retroreflective target as shown in Fig. 3.42. It is possible to
get about 10 times the range with a 3-in.-diameter retroreflector than is possible with a diffuse white reflector. For example,
a 4-in. fiber optic proximity device would likely sense a 3-in.
retroreflective target out to a range of 3 ft. However, one must
also understand that objects crossing within the proximity range
will not break the beam because there will be a smooth transition
from reflex target signal to proximity target signal reception. It
is good practice to adjust the sensor gain so that the proximity
detection range is at most only one half the range of the closest
position the object will ever pass. In some circumstances reflex
operation can also be accomplished with lens attachments. Here
again the potential proximity detection is a concern. In addition,
if the excess gain of the sensor is too high, it will detect the reflection from the back side of the lens and will "latch up." Units
with much greater than 1 in. of proximity performance are likely
to have difficulties performing well in reflex with an attached
lens. Reflex operation using a bifurcated cable and lens attachment should be attempted cautiously. If your application is just
on the ragged edge of latching up, a small dot of flat black paint
can be applied to the center of the back side of the lens to reduce the weak specular reflection it produces.
    Figure 3.43 shows the relative performance of proximity, un-
lensed reflex and lensed reflex performance for a bifurcated fiber optic cable. The shape of the reflex excess gain curves
demonstrate a flat, or zero-slope, curve close to the sensor.
This is the result of the intimately mixed source and detector
fibers, their overlapped fields of view, and the narrow angular

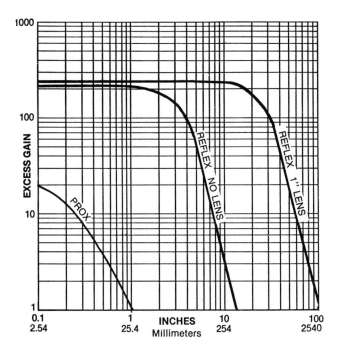

Figure 3.43 Relative performance of proximity, unlensed reflex, and lensed reflex performance for a bifurcated fiber optic cable.

retroreflection. With a 1-in. lens adapter, two effects occur at about 3 ft of range. First, the beam diameter grows larger than the 3-in. reflector, causing some signal to spill over the reflector and be lost. Second, the conical return angle of the retro-reflected light grows larger than the lens, causing some signal to spill over the lens and be lost. Together, these factors cause the excess gain curve to take a downward plunge proportional to the fourth power of range to produce a -4 slope. These factors come into play a little bit differently for each of the fiber optic reflex techniques to form its respective excess gain curve.

3.4.4 Fiber Cable Construction

There are principally three common constructions of fiber cables available in the photoelectric market. The most popular in heavy industry is the flexible armored stainless steel sheath glass fiber

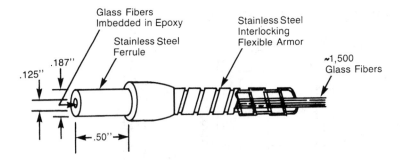

Figure 3.44 Construction of stainless steel armored glass fiber bundle cable.

bundle shown in Fig. 3.44. Over 1000 individual fibers about 0.003 in. in diameter are assembled into a stainless steel sheath and ferrule tip. Within the tip, the fibers are embedded in an epoxy that when cured holds them rigidly. The end of the ferrule is trimmed and optically polished to provide uniformly and squarely terminated fibers. The operating temperature range of this construction is limited by the type of fibers and epoxy used. Standard ratings are typically up to 450°F; however, cables can be constructed on special order to withstand temperatures as high as 1200°F. The interlocked stainless steel spiral sheath provides excellent mechanical protection of the fibers. Although the sheath itself is not waterproof, the individual fibers are not optically affected by liquid contaminants that may enter the cable since the fibers have their own cladding to provide total internal reflection within the fiber. The epoxy in the ferrule prevents any contamination from entering the photoelectric control or attached lens tip. A slightly lower cost construction that uses a steel-reinforced PVC sheath is shown in Fig. 3.45. It uses the same ferrule design and fiber bundle described for the stainless steel sheath variety. Due primarily to the PVC sheath, it is temperature limited to about 200°F in continuous operation. Both of these designs use the industry standard 0.50-in.-long 0.187-in.-diameter ferrule interface to the photoelectric. The standard fiber bundle diameter for this style of cable is 0.125 in.

The most recent construction style to enter the market is the plastic monofiber cable shown in Fig. 3.46. These low-cost fibers were originally designed for use in automobiles but are

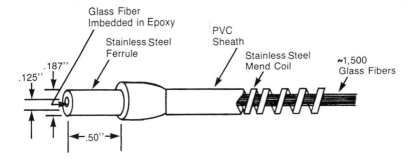

Figure 3.45 Construction of PVC-sheathed-steel-spiral-reinforced glass fiber cable.

becoming quite popular with the lower cost miniature photoelectric sensors available today. The most common size has a 0.040-in.-diameter fiber with 0.085-in. PVC jacket outside diameter. Plastic fiber cables are limited to about 160°F in constant operation. Plastic fibers are much less susceptible to breakage than are the fragile glass fiber bundles and can therefore be pulled in conduit with much less risk of damage. Table 3.1 lists the physical properties and limitations of both glass and plastic fiber cables.

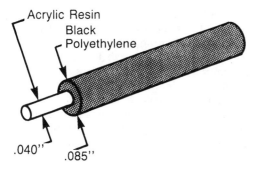

Figure 3.46 Construction of PVC-sheathed-plastic monofiber cable.

Table 3.1 Physical Properties of Photoelectric Fiber Optic Cables

	Standard	Stainless	High temperature	Plastic
Materials	Glass Epoxy Stainless Brass PVC	Glass Epoxy Stainless Brass	Glass Epoxy Stainless Brass	Acrylic Polyethylene
Temperature range	-40 to 105°C	-140 to 230°C	-140 to 480°C	-30 to 70°C
Bend radius	0.5 in.	1.0 in.	1.0 in.	0.4 in.[a] 0.2 in.[b]
Tensile stress yield point	c	c	c	22 lb
Loss/ft at 660 nm	10%	10%	10%	1%

[a]10% transmission loss.
[b]No permanent damage.
[c]Not specified due to construction variability, but generally susceptible to breakage of individual fibers over a few pounds of pull force.

     The light-transmission properties of a fiber optic cable de-
pend on the fiber materials and the color or wavelength of light
being transmitted. The properties of standard glass and plastic
fiber cables were shown in two different forms in Figs. 2.35 and
2.42. In addition to these, many other fiber materials are available from fiber cable manufacturers, including fibers that have
particularly good transmission in the ultraviolet region and fibers which are specifically designed for high temperature operation. In any case, light is attenuated by a fixed percentage for every foot of travel within the cable. The amount of
attenuation will vary from one cable style to another. The
product data sheet or manufacturer should be consulted for
exact information regarding signal attenuation for special length
cables.

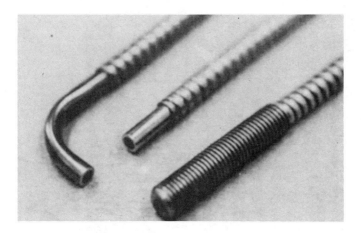

Figure 3.47 The three most common glass fiber bundle tips with 0.125-in.-diameter sensing area.

3.4.5 Fiber Cable Tips

One of the properties of fiber optics that gives them such utility is the great variety of available fiber cable tips. Custom-designed tips are relatively easily constructed to provide almost any sensing shape. The key to success in many sensing applications is the ability to control the sensing area shape to match the object size or shape, or the ability to get the sensor into a particularly confining location. Many odd-shaped fiber tips are readily available as standard product. Figures 3.47, 3.48, and 3.49 are examples of only a few of the many varieties available as standard product.

The most common type of fiber tips are those of Fig. 3.47. These provide a 0.125-in.-diameter sensing shape with various mounting styles. The smooth straight and right-angle tips are typically inserted into a 3/16-in.-diameter hole and held with a setscrew. The standard threaded tip is 5/16 in.-24 threaded brass. Figure 3.48 shows miniature tips with glass fiber bundle diameters of 0.045 in. One of the more interesting problem-solver tips available is a 3-in.-long probe made of soft stainless steel that can be bent to conform to special mounting requirements. These fiber tips are designed for detection of very small objects, small parts assembly, and for very precise position detection of a passing object. Because of their small size, they

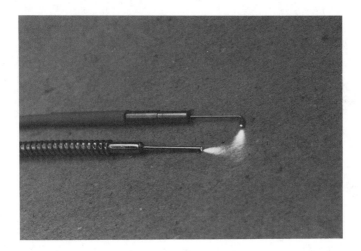

Figure 3.48 Straight and right-angle miniature probe tips.

have significantly reduced excess gain relative to their 0.125-in.-diameter brothers. In thru-beam or proximity, the excess gain of the 0.045-in.-diameter tips would be approximately 60 times less and have a sensing range only one-eighth as far. However, this is not usually of consequence when dealing with very tiny objects. The paddle-shaped tips in Fig. 3.49 are useful for detecting objects with poorly controlled positions, such as falling objects, wires, or web defects. Paddle tips are also useful in edge or position guiding when used with analog output sensors.

The lens tip adapters shown in Fig. 3.50 provide additional optical gain or range for fiber optic sensors, as described in previous sections. Lens tips are generally made to screw onto the 5/16-in. 24-threaded tips. Lenses are best applied to thru-beam applications. Lenses can be used with proximity tips to extend proximity or reflex performance, but may detect the reflection off the back side of the lens if the sensor excess gain is too high, resulting in a latched-up sensor. This can be remedied by turning down the sensor gain or by putting a small dot of flat black paint in the center of the lens on the inside to eliminate the specular reflection from the glass surface.

Liquid-level sensing may be accomplished with a prismatic probe as shown in Fig. 3.51. This probe operates on the principle of total internal reflection. When air surrounds the prism

Figure 3.49 Paddle-style fiber optic tips.

at the end of the probe, light is totally reflected back up the
probe to the detector fibers. When liquid contacts the end, the
critical angle is reduced and the light escapes into the liquid.
These probes are available in both glass and plastic to provide
resistance to different chemicals.

3.5 SCANNERS

An electronic scanner is a device that gathers information through
sensing many points along a defined scan path. A scanner is
the combination of sensing and motion. Motion may be the result
of moving the sensor, sweeping the beam with a moving lens or
mirror, or electronically synthesizing motion by sequentially step-
ping through an array of sensors. Using this strict definition,
in this section we describe photoelectric devices that provide
both the sensing and motion aspects of scanning. Some optical

Figure 3.50 Lens tip attachments for fiber optics.

scanners will be mentioned only briefly, as they fall into other, more sophisticated categories of optical sensing, such as optical instrumentation or machine vision.

3.5.1 Diameter Scanners

Diameter measurement is important in the production of many basic materials. Three scanning methods used in different industries portray a difference in requirements and solutions.

The forest products industry is concerned with the primary breakdown of logs into dimension lumber for building construction. The diameter and shape of each log is unique. The diameter of incoming logs in most modern sawmills is measured by a log diameter scanner in two dimensions to create a computer

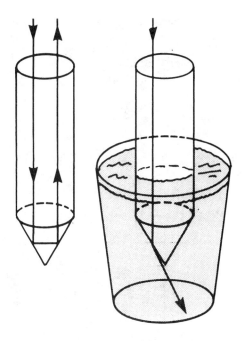

Figure 3.51 (Left) Liquid-level sensor tip; (right) When liquid contacts the prism at the end, total internal reflection ceases as the light is lost into the liquid.

model of the log profile. The computer is then able to determine a cutting pattern to produce the highest possible yield from the log. Most log diameter scanners are constructed with a linear array of infrared source LEDs in one enclosure and a linear array of detectors in a second enclosure. Figure 3.52 shows a section of the source LED array of a log diameter scanner designed for variable-length assembly. The physical construction of the detector array is very similar in appearance. In operation, a single source and detector pair are turned on sequentially, one at a time, for a few microseconds each. The receiver circuitry synchronously determines if the source light was detected and stores the result for processing or transmission. The sequential scanning of the photoelements is referred to as time-division multiplexing. This means multiple signals that would normally interfere with each other can be received independently by giving each a specific slot of time in the electronic

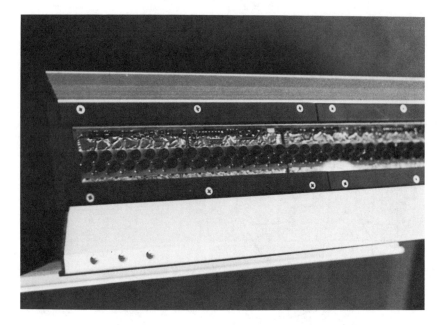

Figure 3.52 LED array section of a log diameter scanner.

scan sequence, as shown in Fig. 3.53. As with other thru-beam
systems, each source and detector pair form an effective beam
the shape of a small rod that joins the lens of each source and
detector element. Figure 3.54 shows how the individual beams
of the scanner are interrupted by the log as it passes between
the scanner heads. This construction provides 0.1-in. resolution and scan rates as high as 6 in./ms. A total scanning system consists of optical scanner heads, power supply, scan controller, and optional high-speed computer interface, as shown
in Fig. 3.55. This construction method provides the gross resolution required for logs, uses thru-beam sensing, which is most
tolerant of contamination, and inherently eliminates perspective
scaling distortion errors by use of parallel beams.

Small-diameter extruded or pulled materials such as wire and
pipe require much higher resolution measurements to maintain
production quality standards. This measurement is often made
with a parallel-beam scanning laser, as depicted in Fig. 3.56.
In this device a thin laser beam is directed toward a rotating mirror at the focus of a lens. The light is caused to scan across

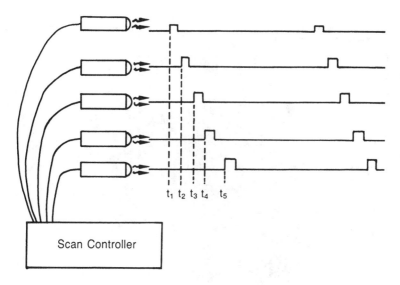

t_1 t_2 t_3 t_4 t_5

Scan Controller

Figure 3.53 Time-division multiplexing assigns a slot of time for each sensor in order to avoid crosstalk between sensors.

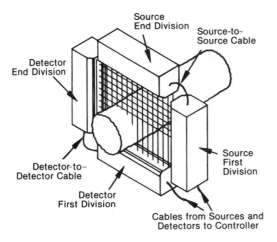

Figure 3.54 Log diameter scanner heads transmit parallel beams of infrared light. Detector heads sense the shadow diameter. (Courtesy of Opcon, Inc.)

Figure 3.55 Log scanner system includes scan heads, scan controller, and optional high-speed computer interface. (Courtesy of Opcon, Inc.)

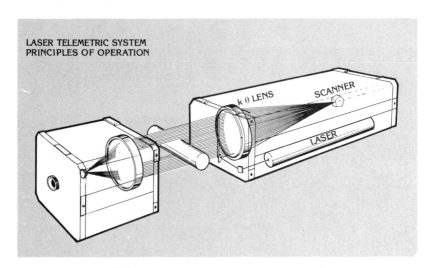

Figure 3.56 Scanning laser beam diameter measurement. (Courtesy of Zygo Corporation.)

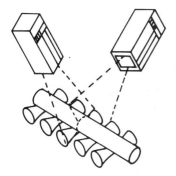

Figure 3.57 Scanned CCD array cameras determine hot pipe diameter and eccentricity. (Courtesy of Opcon, Inc.)

the face of the lens by the rotating mirror. The lens bends the
radially scanning light path into a parallel scanning light beam
directed toward the target and detector. Light is collected by
a second lens on the far side of the target and focused onto the
detector. The signal from the detector will contain a dip in
strength when the laser beam is blocked by the target. The exact point in time and duration of light blockage corresponds to
the target position and diameter. The raw signal of this instru-
ment is processed by an integral microcomputer to convert it into a useful form.

Hot-rolled steel pipe is measured in two axes to determine
both the diameter and the eccentricity of the pipe. This measurement is accomplished with a pair of scanned linear array
cameras mounted at 90° to one another. The mounting arrangement is shown in Fig. 3.57. The CCD array cameras are quite
sensitive to the infrared light emitted by the hot pipe and use
it directly as the source of illumination. Perspective distortion
is removed by a microcomputer using the information from the
other camera to infer the distance to the pipe. The scanning of
the CCD array detectors is controlled by the microcomputer in
this one-dimensional line-scan machine vision system.

3.5.2 Safety Light Curtains

Safety light curtains are intended to provide a photoelectric
safety interlock to prevent operation of potentially dangerous
equipment when human beings or other objects are not at safe

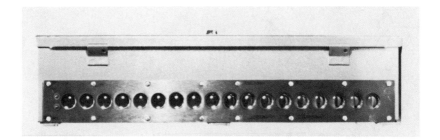

Figure 3.58 Light safety curtain constructed with multiple scanned thru-beam sensors. (Courtesy of Dolan-Jenner Industries, Inc.)

distances. Concern for human safety has increased significantly in the past decade and has prompted the Occupational Safety and Health Administration (OSHA) and the American National Standards Institute (ANSI) to set standards for the reliability and construction of these devices. Safety light curtains generally operate on one of two principles, both similar in technology to diameter scanners. Most commonly they are constructed with a scanned array of thru-beam sensors with lens elements at about 3/4 in. spacing, as shown in Fig. 3.58. Individual source/detector pairs are turned on and off sequentially to produce a parallel scanning beam effect. A second construction method shown in Fig. 3.59 uses a single reflex sensor looking into a rotating mirror that directs the transmitted beam into a parabolic mirror to create a scanning parallel beam that traverses the height of the enclosure. The retroreflected light returns along the same path until part of it is split off by a partially silvered mirror and directed onto the detector. Safety light curtains are capable of being set to ignore specified regions where blockage may occur due to machine structure or infeed stock.

Because safety light curtains involve human safety, which has implications of potential liability suits, many photoelectric manufacturers will not participate in this market. When such equipment is used, the manufacturer heavily stresses regular testing and maintenance according to the instructions provided. Local, state, and federal codes, regulations, and laws must be complied with in the installation and use of this equipment.

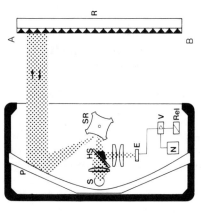

Light Curtain

Same principle as the reflected beam with the addition of a rotating mirror wheel SR and a parabolic reflector P, producing a parallel scanning beam. This beam moves at high speed from A to B, thus forming a "light curtain". Interruption of this light curtain is detected and signalled by the relay Rel.

S = light source
HS = semi-transparent mirror
SR = mirror wheel
P = parabolic reflector
R = reflector
V = amplifier
Rel = relay
N = power supply
E = receiver (photocell)

Figure 3.59 Light safety curtain constructed with rotating beam deflector, parabolic mirror, and single reflex sensor. (Courtesy of Sick Optik-Elektronik, Inc.)

3.5.3 Multiplexed Remote Sensor Scanners

Multiplexed remote sensor scanners offer flexibility in application when an array of sensors are needed. Multiplexing means that two or more signals are carried in the same medium (electrical wire, radio waves, light, etc.) without interference between them. Photoelectric scanners almost always use a technique called time-division multiplexing to prevent sensor crosstalk. These scanners are particularly useful when many thru-beam sensors must be stacked together or many proximity sensors are sensing in the same general area. Most multiplexed sensors have a separate scan controller, as do those in Fig. 3.60. In the variety shown in Fig. 3.61, the individual sensors create synchronizing signals that enable them to communicate with each other and multiplex themselves without a separate scan controller.

3.5.4 Other Scanners

There are numerous other photoelectric-related scanners marketed that will be mentioned only briefly. These include bar code scanners, optical character recognition (OCR), web defect scanners, machine vision, and others. Only a short description of these will be offered, as they are not the intended subject matter of this book and represent separate branches in the industrial use of optoelectronics technology.

Bar code scanning has become the norm for identification of groceries at the checkout stand, inventory identification and transaction, and material routing control in automated warehouses. Bar code scanners are an important branch of photoelectrics in a very specialized niche. Scanning is accomplished either via a moving helium-neon laser beam producing a small red spot that traverses the bar code, or via a wand with an optical tip that is stroked across the bar code. Timing of the changing intensity pattern of reflected light as it crosses over the bars is analyzed with a microcomputer to reconstruct the bar pattern digitally in memory. The bar pattern is then deciphered to produce the numerical code it represents.

Optical character recognition (OCR) has long been the ideological scanner for printed information since it requires no additional printed codes to make the item machine readable and adds no aesthetically unpleasing printing to product packages and labels. OCR has turned out to be a significant technical challenge that has only recently become reasonably reliable for

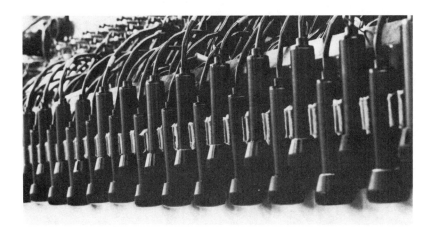

(a)

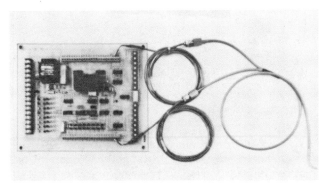

(b)

Figure 3.60 Remote sensor multiplexed scanners with separate scan controller. [(a) Courtesy of Opcon, Inc.; (b) courtesy of Dolan-Jenner Industries, Inc.]

defined sets of fixed type fonts expected by the scanner. OCR is currently successful in applications such as bank check number reading, typewritten text computer input, lot and date code verification on pharmaceutical labels, and reading of serial numbers or part numbers on assembly line parts. OCR has been

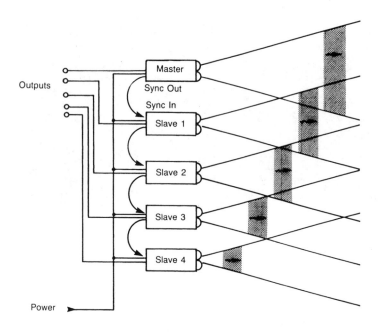

Figure 3.61 Self-multiplexing sensors scan without need for separate scan controller. (Courtesy of Opcon, Inc.)

only marginally successful in reading price codes at the checkout
stand, has had insufficient success in postal applications and has
had no success at reading the variable style and placement of type
on product labels. Similar to bar code reading, the OCR scanner
produces an image of the text in the microcomputer memory. The
deciphering program must then identify potential character blobs
and match them with known character shapes with some degree of
certainty. Because of the similarity and complexity of some char-
acters, this has been a difficult task for our character set and at
present an impossible task for the Eastern kanji character set.

Web defect scanners are used in the production of continuous
sheet material to inspect for blemishes and voids that degrade
product quality. Production speeds of webs may vary from a
few inches per second to hundreds of inches per second. Webs
may be inspected for uniformity of raw stock, uniformity of coating, or registration and quality of a printed pattern. Slower webs
are often inspected with a linear scanned array camera. Faster

webs being inspected for uniformity, blemishes, and voids are
generally scanned by very expensive flying spot scanners. These
scanners traverse a laser spot across the web at speeds up to
14,000 ft/s to attain sufficient coverage. Detection of a flaw is
made by sensing changes in reflectivity or transmissivity.

Machine vision systems are composed of one- and two-dimensional scanned array cameras which produce an analog video output voltage that is digitized and loaded into computer memory.
Algorithms within the user's application program sort through
this digitized numerical representation of the image to determine
size, location, orientation, quality, or other aspects of interest.
Photoelectric sensors are generally used to provide a trigger signal to the machine vision system when a passing object to be inspected is in position. Machine vision is currently a rapid-growth
market fragmented into many specialized areas. As the cost of
low-end machine vision products drops, they will probably com-
pete well with some of the more complex photoelectric sensor and
sensor system products.

4

Photoelectric Output Interfaces

What you know I need to hear
Words you use might be clear
We talk the same, but we don't
I want to know, but I won't

Witatschimolsin

In this chapter we describe various photoelectric sensor and control output interfaces, what they are, principles of operation, basic terminology, ratings, and limitations. Output device failure is the most common photoelectric failure mode. Installation miswiring, output overload, and relay contact wear account for the majority of these failures. Additionally, application failures often result from inappropriate output device selection. Output devices for both electronic signaling and heavy-duty power control will be discussed. In this chapter we present the information required for understanding these output devices and applying them reliably.

4.1 ELECTROMECHANICAL RELAYS

For many years the electromechanical relay was the dominant output device for photoelectric controls. Even today relays constitute a significant portion of output device sales. The relay's capability to switch large currents at high voltages allows it to control directly other major equipment, such as motors, brakes, ejectors, and heating elements. Relays are capable of switching AC or DC power, they provide isolation between the load circuit

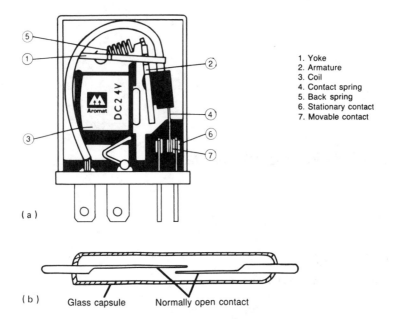

1. Yoke
2. Armature
3. Coil
4. Contact spring
5. Back spring
6. Stationary contact
7. Movable contact

(a)

(b)   Glass capsule   Normally open contact

Figure 4.1 (a) Armature relay mechanism; (b) reed relay mechanism removed from its surrounding coil. (Courtesy of Aromat Corporation.)

and the photoelectric amplifier, they produce the least heat when switching moderate to large currents, and they are familiar to virtually all factory personnel. Relays are not without problems. They are subject to mechanical wear, they produce electromagnetic radiation, their contacts are eroded by electrical arcing, their response time is relatively slow, and their contact bounce can produce erratic results in electronic counters.

4.1.1 Construction

The most common relay construction used in photoelectric controls is shown in Fig. 4.1a. An energized coil of wire magnetically attracts a moving armature on which electrical contacts are mounted. As the armature moves toward the magnetic coil, electrical contacts are made and/or broken. The reed relay, shown in Fig. 4.1b, is composed of two plated iron reeds sealed in a

glass capsule that is surrounded by a coil of wire. When the coil
is energized, it creates a magnetic field that causes the reeds to
attract and stick to one another, closing the contacts. The armature relay is most commonly used in photoelectric controls and
is almost always socketed for easy replacement. Energizing coils
are designed and specified for particular operating voltages.
Photoelectric manufacturers specify and supply relays with coil
operating voltages compatible with the photoelectric internal circuitry. Contact arrangement, contact plating, and case sealing
are also important features for compatibility and maximum operating life. Replacement relays must have the same physical and
electrical characteristics as the original to ensure proper operation.

4.1.2 Contact Styles

There are dozens of contact arrangement forms available from relay manufacturers. However, photoelectric controls generally
use one of the three forms shown in Fig. 4.2. A relay can be
thought of as a common switch terminal that may be connected
to one or more output terminals, depending on the position of
a magnetically moved switch conductor. An output terminal is
called normally closed (NC) if it is electrically connected to the

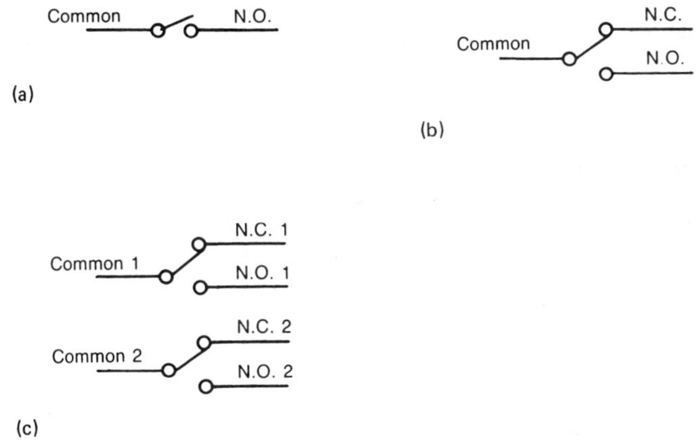

Figure 4.2 (a) Single-pole single-throw relay (SPST), form 1A;
(b) single-pole double-throw relay (SPDT), form 1C; (c) double-
pole double-throw relay (DPDT), form 2C.

common terminal only when the relay coil is not energized. Conversely, it is called normally open (NO) if it is not electrically connected to the common terminal until the relay coil is energized. A single relay may have many sets of contacts with any combination of normally open and normally closed contacts. A relay with a single common terminal and an NO and an NC terminal is said to have a single pole (one common) and double throw (either NO or NC), and in shorthand notation is called a SPDT relay. A second convention for describing the contact arrangement is given in Table 4.1. This convention describes the contact arrangement with a letter designator and numeric multiple. For example, form 1C is a SPDT; form 2C is two form 1C switches, or alternatively, a DPDT arrangement. Many other more complex forms are designated and manufactured. More information on contact arrangements is available from relay manufacturers' literature (e.g., *Relay Technology* (Mountainside, N.J.: Aromat, Inc., 1983), p. 9].

There are four primary contact styles shown in Fig. 4.3: reed, crossbar, bifurcated, and single button. The simple reed contact is found in low-level current-switching reed relays. It has low contact pressure and a single point contact. The crossbar contact style provides much greater contact force per unit contact area, which improves the reliability of relays with low armature force. Bifurcated contacts provide improved reliability through redundancy, reduction in arcing time, and less contact bounce. Single button contacts are quite sturdy and well suited to heavy-duty circuits.

Table 4.1 Common Contact Arrangement and Form Designations

Arrangement	Form
SPST NO	1A
SPST NC	1B
SPDT	1C
DPST NO	2A
DPST NC	2B
DPDT	2C

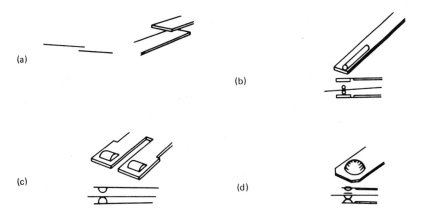

Figure 4.3 Contact styles: (a) reed; (b) crossbar; (c) bifurcated; (d) button. (Courtesy of Aromat Corporation.)

4.1.3 Contact Materials

One of the most important considerations in relay selection for
application reliability is the choice of contact material. When
contacts make and break a circuit, high currents and showering
arcs can cause metallurgical and chemical changes on the contact
surfaces, rendering them unreliable. Arcing can cause contact
welding, pitting, and burning. The occurrence of oxide and
sulfide buildup can result in the contacts failing to close electrically even though they have physically contacted one another.
To establish electrical contact, the oxide film must be overcome.
Arcing can provide a means of breaking through the film. Materials that work well in high-current applications are often unreliable in low-current applications. As a result, relay manufacturers have developed, and recommend, different contact materials for different applications. Before selecting a relay contact material, the application load should be reviewed to determine inrush current at contact make, continuous current, and
the presence of inductive kick at contact break. Photoelectric
manufacturers generally supply relays designed to handle high-
current and high-voltage loads. These relays are not suitable
for logic-level interfaces and will probably fail early. Logic-
level interfaces are best handled by solid-state relays or gold-
plated relay contacts. A review of contact material properties
will provide insight to improved relay application reliability.

Fine silver and silver-cadmium-oxide contacts are quite common in power relays. They should be used in circuits with over 12 V and 1 to 10 A or more. The oxides and silver films cannot be broken through reliably with less than 12 V. Silver and silver alloy contacts are resistant to oxidation but are plagued by sulfidation. Silver sulfide films must be broken through with contact pressure, wiping action, or burned through with controlled arcing. Silver-cadmium-oxide handles slightly larger currents due to its greater resistance to welding and pitting. Some manufacturers produce sealed relays that prevent sulfidation in severe environments.

Gold or gold alloy contacts are used primarily for low-power switching. Gold is very resistant to chemical attack and oxide formation. Gold flash contacts are easily burned through by electrical arcing in just a few operational cycles. Arcing will not occur when circuit voltage is below 12 V or when circuit current (including inrush current) is below 0.5 A. Gold is soft and may stick or cold weld at higher currents. Gold alloys reduce the cold-weld problem and are generally less expensive.

Rhodium is used primarily on reed relay contacts and is useful for switching 10 mA to over 1.0 A. Rhodium is quite resistant to oxidation and corrosion, does not cold weld, and has stable contact resistance.

Tungsten contacts are used in applications where high inrush currents or very high voltages are required. Tungsten has a very high melting temperature and is highly resistant to contact errosion. Tungsten easily forms oxide films and is not recommended for applications under 24 V. Tungsten contacts are popular for motor starting and incandescent lamp control applications.

Mercury wetted contacts are most common in reed relays. The liquid mercury clings to the metal surfaces of the contacts. Because it is a liquid, there is no issue with contact material transfer or oxide buildup. The surface renews itself each time the contacts operate. Mercury wetted contacts exhibit very stable contact resistance and bounce-free operation. They are capable of switching less than 1 mA to over 1 A.

4.1.4 Contact Protection

Under certain conditions of current and voltage, a disruptive discharge may occur during contact breaking that causes contact damage and produces electromagnetic interference. These phenomena are most severe when an inductive load circuit is opened

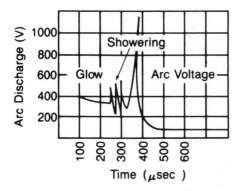

Figure 4.4 Glow and arc discharge voltages as measured at the relay contacts when opening an inductive load circuit. (Courtesy of Aromat Corporation.)

but are not limited entirely to inductive loads. Figure 4.4 shows three separate phenomena that may occur. The glow discharge occurs at currents under 0.4 A only when the breaking voltage is above approximately 300 V. A showering arc discharge is an oscillatory discharge with characteristics determined by the inductance and capacitance in the circuit. This phenomenon is particularly notorious for generating conducted and radiated electromagnetic interference. The arc discharge can occur when contact current is larger than about 0.4 A and contact voltage exceeds 12 V for silver and 15 V for gold. Figure 4.5 shows in more detail the minimum arc starting voltage versus current relationship for common contact materials. The concentration of energy in an arc discharge is very high and can cause localized temperatures to rise as high as 2000 to 10,000°C. Such high temperatures can easily melt contact surface material, causing pitting, holes in gold flash plating, and material transfer from one contact to another. A second area of concern is contact closure on circuits that exhibit very high inrush currents. These include capacitive loads, incandescent lamps, and motors. Each of these can have inrush currents as high as 10 to 40 times the normal continuous current for that load. These high currents can cause localized melting of the contact surfaces as the tremendous currents flow through tiny contact points. The results can be sticking or cold-welded contacts, or overheated contacts if the relay is cycled too rapidly.

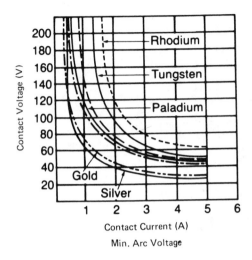

Figure 4.5 Minimum arc starting voltage and current. (Courtesy of Aromat Corporation.)

One of our principal objectives in the application of photoelectric controls is reliability. It is possible to increase the life and reliability of the relay by reducing or eliminating destructive discharges present on opening or closing the contacts and generally not overloading the contacts. Energy is stored in the magnetic field created by the electrical current that flows through the coil of inductive loads, such as solenoids, contactors, or motors. When the coil current is interrupted by the opening of the relay contacts, the coil generates a very large kick voltage in response. The magnitude of the kick voltage is proportional to the rate of change of current flowing in the coil. The opening of the contacts attempts to force the current flow to stop instantaneously, generating tremendous kick voltages. As the kick voltage is generated, however, it finds two ways to cause current to continue to flow. First, parasitic circuit capacitance is charged by the rapidly increasing kick voltage. The current that flows into the parasitic capacitance limits the rate of rise of the inductive kick voltage. Second, as the kick voltage increases, it will eventually reach the breakdown voltage of the gap between the contacts, striking an arc and restarting the current flow until the energy stored in the magnetic field is depleted. The

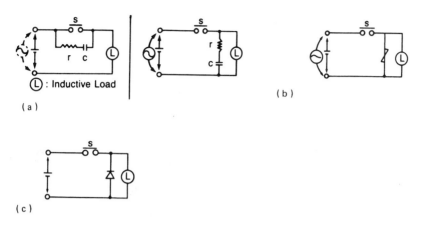

(L) : Inductive Load

(a)

(b)

(c)

Figure 4.6 Solutions for relay contact protection include (a) RC snubber for AC or DC, (b) varistor clamp for AC or DC, and (c) diode clamp for DC circuits only.

three most common ways of reducing or eliminating the inductive kick are with RC snubbers, varistor clamping, and diode clamping, as illustrated in Fig. 4.6. An RC snubber consists of a series resistor and capacitor in series placed across the relay contacts. The capacitor reduces the rate of change in inductor current when the contacts open by absorbing the current generated as the kick voltage increases. By reducing the rate of change in current flow through the inductive load, the inductive kick voltage is reduced. Figure 4.7a shows how the snubber reduces the inductive generated kick voltage when an RC snubber is added to the circuit. RC snubbers are used primarily on AC circuits, although they may also be used successfully on DC circuits. Potter & Brumfield suggest choosing resistor and capacitor values according to the nomogram in Fig. 4.8. To use the nomogram, a straightedge is laid between the load voltage and load current values. The resistor and capacitor values are then read off the other scales directly. The voltage rating of the capacitor should be compatible with the circuit supply voltage, and a resistor wattage of 1/2 W will handle virtually all circumstances. The second contact protection method recommended for AC circuits is the varistor clamp. A varistor is a resistor that is sharply dependent on the voltage across its terminal. At low voltages it has very high resistance. Above

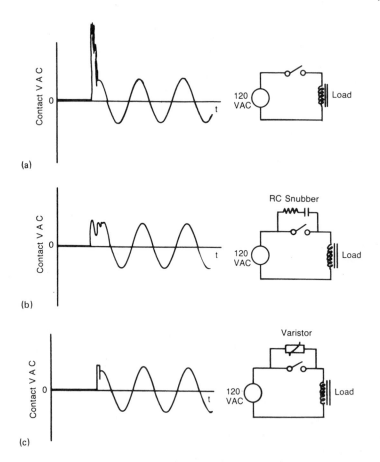

Figure 4.7 Ac circuit inductive load relay contact voltage when opening the contacts for (a) unprotected, (b) RC snubbed, and (c) varistor clamped.

some high knee voltage, its resistance becomes very low and it conducts current quite well. A varistor across the relay contacts will allow the inductive kick voltage to rise only as high as the knee voltage; then it will dissipate the stored inductive energy as current flows through it at the clamped knee voltage. Circuits of 120 VAC can be clamped at about 200 V and still protect the contacts from the glow discharge. In DC inductive load circuits, the diode clamp method shown in Fig. 4.6c is recommended.

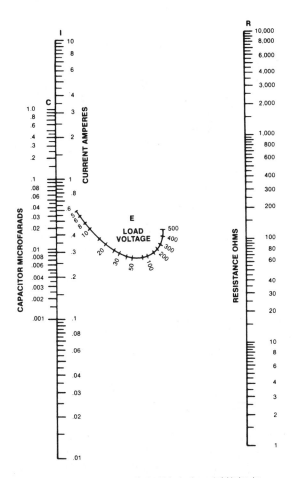

Resistance (R) in ohms is obtained using a straightedge. Locate 1.0 amperes (I) on the left-side scale and 50 volts (E) on the center scale. Place the straightedge on these points. The junction of the straightedge and the right-side scale determines R. In this example R is equal to 5.0 ohms.

Figure 4.8 Nomogram for determining resistor and capacitor values for RC snubbers in AC inductive load circuits. (Courtesy of Potter & Brumfield, A Siemens Company)

The diode clamp is placed across the inductor with its cathode tied to the positive supply. As the inductive kick voltage on the relay contacts rise, the diode will eventually become forward

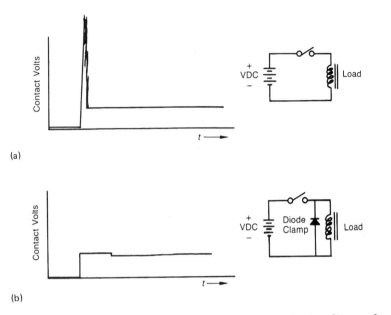

Figure 4.9 DC circuit inductive load relay contact voltage when (a) unprotected, and (b) diode clamped.

biased and clamp the maximum inductive kick to the supply as shown in Fig. 4.9b. The clamp diode should be rated to handle the full load current and withstand twice the supply voltage. Clamping and snubbing are effective methods for controlling low-current glow discharge triggered by the presence of high voltage across the contacts when opening. High-current arc discharge may be triggered at relatively low voltages even without an inductive kick and should use the RC snubber method. These techniques significantly reduce electrical noise and can significantly increase the life expectancy of relay contacts in inductive circuits.

4.1.5 Rated Life

The rated life of a relay depends on how it is used or abused. The following discussion will center around electrical contact wear since it is a more significant limitation than mechanical failure. The minimum expected mechanical life of medium-sized

Change in contact resistance

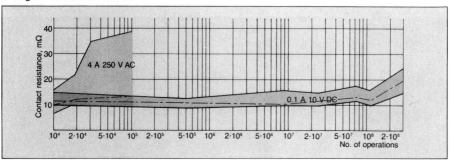

Figure 4.10 Relay contact resistance change over life of contacts.
(Courtesy of Aromat Corporation.)

relays that handle 10 to 20A is generally specified for 10 million
operations, miniature relays for 100 million operations, and reed
relays for 1 billion operations.

The life of a relay contact is generally defined as the point at
which the contact resistance has increased beyond a set level.
Often this level is set at about 0.10 Ω. Contact resistance greater
than this value may produce unacceptable heating levels on the
contact, further hastening contact deterioration. Higher currents
and voltages shorten the useful life of relay contacts. Figure 4.10
shows that the difference in useful life can be of an order of magnitude or more. Gold-flashed contacts perform very well over
millions of operations with low current and voltage. In Fig. 4.11
we see the effects on contact life for different voltages, currents,
and power factor loads.

Phi (ϕ) is the phase angle between the current and load voltage. Cos ϕ is called the power factor. For cos ϕ = 1, the load is
resistive; for cos ϕ = 0.4, the load is inductive. These life curves
do not indicate that the relay will stop functioning after the specified number of cycles. The curves indicate that the contacts have
started to degrade significantly. Fortunately, most applications
can operate sufficiently with very poor condition contacts. A set
of relay contacts is rated for the current it switches, not the continuous current. Motors, lamps, and solenoids all have inrush
currents at turn-on many times that of their steady-state current.
These currents should be known before specifying a relay. Because parallel contacts on a form C relay do not make contact at
the same instant, it is generally of little value to tie them together

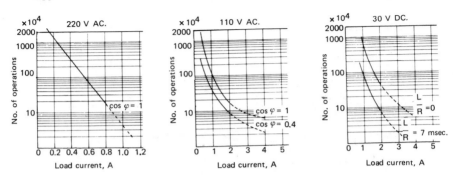

Figure 4.11 Relay life curve for 2-A bifurcated silver contacts. (Courtesy of Aromat Corporation.)

electrically to extend contact life. A parameter often poorly estimated without calculation is the number of days a relay can be expected to operate reliably. If a relay operates once every second, it will operate 2 million times in one month! For an expected relay life of 100,000 operations, replacement may be required daily! Finally, the environment can play a significant role in contact life when even small quantities of chemicals are in the air. Figure 4.12 shows how silver contacts rapidly sulfidate in the presence of H_2S gas, a chemical quite common to pulp and

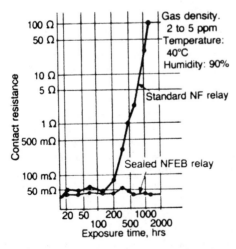

Figure 4.12 H_2S test for sulfidation of silver contacts. (Courtesy of Aromat Corporation.)

paper processing and many other industries. In just a few days the contacts may be so well corroded that they will no longer switch low-level signals. Where available, a sealed relay can prevent corrosion from the environment. Alternatively, controlled arcing can be used to burn through the corrosion film. Check with your photoelectric manufacturer or their relay supplier for alternative relays if you require special contacts to ensure reasonable life expectancy.

4.2 SOLID-STATE RELAYS

Solid-state relays are becoming the AC circuit switching device of choice. In earlier years, solid-state AC switches were not well accepted, for two reasons. Some early implementations of solid-state relays did not function well with inductive loads and left some pioneer users feeling betrayed. Second, solid-state relays did not have that old familiar "click click," which gave the warm, comforting feeling of a working photoelectric. Today's solid-state relays are superior to electromechanical relays in almost every way. Solid-state relays produce much less electrical switching noise, have extremely high reliability and long life, are optically isolated from the control circuit, and are tolerant of very high shock and vibration. Solid-state relays are not perfect switches. Most have an on-state voltage drop of 1 to 2 V; they produce more heat than do relay contacts, which limits their switching current in small photoelectrics; they may require a leakage-current-generating RC snubber to prevent latching on with some inductive loads; some designs do not work in series or parallel with similar switches; and some are not capable of switching DC loads.

4.2.1 Construction

Solid-state relays are constructed of electronic components having no moving parts. Most often the solid-state relay circuit is encapsulated either in a plug-in module, as shown in Fig. 4.13, or as an integral part of the photoelectric control circuitry. A solid-state relay consists of an input circuit, an optocoupler, a power output circuit, and power circuit protection components, as shown in Fig. 4.14. One of the primary properties of a relay is its ability to isolate the control circuit from the output circuit, which may be subject to high voltage and transient voltage spikes, which could easily disrupt or destroy the control circuit. The key to circuit isolation in solid-state relays is the optocoupler,

Figure 4.13 Typical plug-in solid-state relay circuitry prior to encapsulation.

a small device consisting of an LED that shines light through an optically transparent electrical insulator toward a photosensitive semiconductor device that is activated by the light. The semiconductor device may be a phototransistor, photo-SCR, photo-

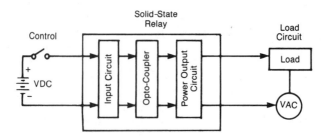

Figure 4.14 Block diagram of a solid-state relay.

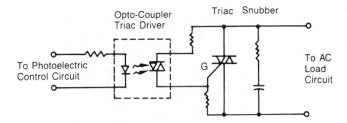

Figure 4.15 Schematic of triac solid-state AC relay.

triac, or photovoltaic FET driver. Each of these devices is suited
to driving specific types of power output circuits. The power output circuit, based around a silicon power semiconductor switch,
has a number of implementations and usually has associated circuit protection components to prevent damage or false triggering
due to transient voltage spikes. The four most common implementations for AC solid-state relays are represented by their equivalent schematics in Figs. 4.15 through 4.19. Each of these circuits
has its advantages and quirks.

4.2.2 Triacs

The most common and lowest-cost solid-state relay implementation
is the representative triac circuit of Fig. 4.15. A triac is a bilateral switch, meaning that it operates from either polarity of
voltage and current. In normal operation, the photoelectric
drives current into the optocoupler, causing the internal LED
to produce light. The light is transmitted through an optically
transparent electrical barrier and strikes the light-sensitive triac
driver. The triac driver switches on and causes current to flow
into the gate terminal of the power triac. A small current flowing through the triac gate terminal triggers conduction of current between its two main terminals. Some low-current triac currents rated less than 50 mA delete the power triac from the circuit and drive the load directly from the triac driver optocoupler.
After the signal at the gate is removed, the triac will remain on
and continue to conduct until the load current falls below a low-
current threshold which is referred to as the holding current.
This will occur once every 8.333 ms in a 60-Hz AC circuit. Because DC circuits to not have cyclic transitions of current through
zero as do AC circuits, DC circuits become latched-on once the triac

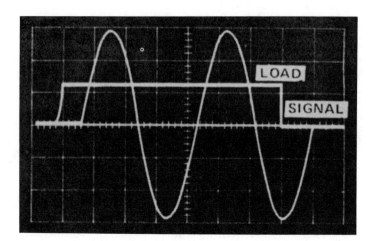

Figure 4.16 DC control signal and AC load voltage oscilloscope traces for zero-crossing solid-state switch. (Courtesy of Crydom Division of International Rectifier Corporation.)

is triggered and cannot be turned off without removing power from the circuit. However, if a latching DC output is required, this can be a clever way to implement it cheaply.

Some optically coupled triac drivers use zero-voltage switching circuitry to turn on the triac only at a zero-voltage crossing. Turning the AC circuit on at zero voltage and turning off at a zero current, as shown in Fig. 4.16, eliminates the line transients and electrical noise normally produced by electromechanical relays and other devices that switch immediately on demand regardless of voltage and current conditions. Zero-voltage and zero-current switching, however, carries the penalty of waiting as long as 8.33 ms for the next zero crossing, adding delay to the output response.

Although zero-current switching eliminates the inductive kick problem when switching inductive loads, there is still a very rapid change in voltage across the triac terminals. Inductive loads have a voltage and current relationship that is out of phase. When the AC current is zero, the AC voltage is at a peak. When the triac turns off at the zero-current crossing, the entire inductor voltage is instantly transferred to the triac terminals. This represents a large change in voltage over time, or large dV/dt. A triac will false trigger if dV/dt is too large. A condition can arise where every time the triac tries to turn off, it is retriggered by the dV/dt for another half cycle. This condition can occur repeatedly every half cycle and result in a latched-on condition. A large dV/dt may also be produced by other devices operating on the same AC power lines. Large dV/dt spikes from external sources may also cause a triac to false trigger, staying on for the remainder of the half cycle. To reduce dV/dt, an RC snubber is generally added to the circuit. The capacitor in the RC snubber increases the voltage slew time. The resistor prevents excess capacitive inrush current and damps ringing from the inductor-capacitor interaction. While solving the dV/dt problem, the RC snubber may create another with the leakage current that will be generated in the off-state as a result of the snubber. A triac rated for 1.0 A inductive will probably require a snubber that passes about 5 mA of leakage current. For very low current loads, leakage current may produce enough voltage on the load that it will seem as if the triac never quite turned off, whereas in fact it has been. There are two solutions to this problem. First, change to an output device that specifies low-enough leakage current. Second, parallel the load with a resistor that will reduce the voltage developed across the load by absorbing some of the leakage current itself. The standard maximum allowed off-state current is 1.7 mA for a programmable controller AC input, the place where this problem most often occurs.

Triacs are capable of handling inrush current many times their specified steady-state current. This is beneficial for driving lamps, motors, solenoids, and contactors. High inrush current repetitively switched can, however, cause the triac to overheat internally and be destroyed. The high inrush currents of these loads are not obvious to many users, but often range up to 30 times the continuous current. For example, switching a 0.5-A lamp three times a second at 50% duty cycle with a 1.0-A-rated triac will overload the triac by about 300%! To estimate if your application will overload the triac, use the expression in Eq. 4.1 to estimate the minimum required continuous current rating. The voltage drop across the triac terminals is about 1 V in the on-state for low currents and increases to about 2 V for high inrush currents. Accidental momentary overloads or short circuits from miswiring can easily destroy the switch. Often no external audible or visual signs will be produced to indicate that it has been destroyed. An external replaceable fuse is a recommendation worthy of consideration:

$$\text{Minimum current rating} = (I_{ir})(t_{ir})(N) + (I_c)(\% \text{ duty}) \qquad (4.1)$$

where

I_{ir} = inrush current, Amps
t_{ir} = inrush time, seconds
N = number of cycles per second
I_c = continuous current, Amps

4.2.3 Inverse-Parallel SCRs

A very similar circuit to the triac circuit is the inverse-parallel
SCR (silicon-controlled rectifier) circuit represented by Fig.
4.17. An SCR is physically constructed and operates similar
to a triac. It requires a gate trigger signal to turn on, and
current to fall below the holding current threshold to turn off,
but conducts current in only one direction. To fully handle
AC loads, two SCRs are connected in inverse-parallel so that
load current may be conducted for either polarity. This results in a more complicated and costly circuit; however, the
advantage of inverse-parallel SCR circuits over triac circuits
is that they are much less sensitive to dV/dt problems. In
most cases an RC snubber is not required, even with most inductive loads. Should a snubber be required, the capacitance
needed will be much smaller and result in much less leakage current. Except for these specific differences, the circuit characteristics are identical to triac circuit characteristics and will not
be repeated here.

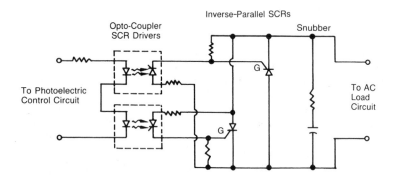

Figure 4.17 Equivalent schematic of inverse-parallel SCR AC
solid-state relay.

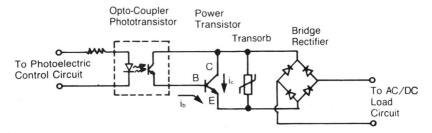

Figure 4.18 Equivalent schematic of NPN transistor AC/DC solid-state relay.

4.2.4 Bridged Transistors

The bridged transistor circuit, represented in Fig. 4.18, is constructed by using a full-wave bridge rectifier to convert current or voltage in the load circuit of either polarity to a single polarity that can be switched by a power transistor. The power transistor conducts current i_c from its collector to emitter terminals proportionally about 30 times greater than the base control current i_b flowing from the base to emitter terminals. The base control current is generated by the phototransistor within the optocoupler when it is illuminated by the optically isolated LED driven by the photoelectric circuit. Transistors are subject to destruction from high-voltage spikes and must be protected. Protection is generally provided by a varistor or transorb clamp placed across the transistor terminals to absorb the energy of high-voltage spikes by becoming low impedance at high voltage.

The bridged transistor design is capable of switching both AC and DC load circuits instantaneously on and off regardless of the phase of voltage or current. The switching action does, however, take tens of microseconds to switch from one state to the other. The transistor is not sensitive to commutating or noise spike dV/dt and cannot be false triggered on for a half line cycle as can a triac or SCR. Leakage current used to power the switch internally is generally less than 100 μA at 120 VAC for a 100-mA output drive rating. The voltage drop across the switch is a hefty 3 V at rated current, due to the 1.5-V drop of the bridge rectifier plus about 1.5 V dropped by the power transistor circuit. Because this circuit drops twice as much voltage at the same currents as triac and SCR circuits, it will generate twice as much heat in operation and limit the minimum package size

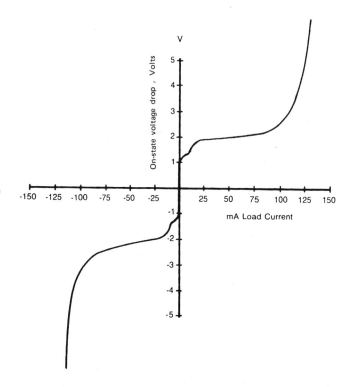

Figure 4.19 Voltage-drop characteristics of a bridged transistor solid-state relay versus load current.

or maximum current rating. The transistor switches are not generally capable of handling large inrush currents. At currents above the rated current, the switch becomes current limited and the output voltage across the switch rises dramatically as shown in Fig. 4.19 and produces a lot of heat within the transistor. The power transistor can easily become overheated and be destroyed if operated repeatedly or continuously beyond its current rating.

4.2.5 Bilateral FETs

The bilateral FET (field-effect transistor) has been hailed by many as the ideal component for solid-state relays. The voltage

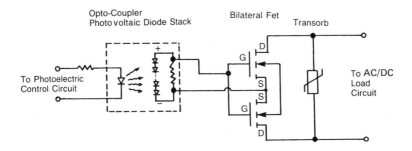

Figure 4.20 Equivalent schematic of bilateral FET AC/DC solid-state relay.

on the gate terminal of a FET controls the conduction resistance
between the drain and source terminals. The bilateral FET circuit represented by Fig. 4.20 has two back-to-back FETs that
control the load voltage, each in one polarity. In normal operation, one FET is properly biased and controls the load while the
other is reverse biased and does not impede the operation of the
controlling FET. When the load voltage polarity reverses, the
FETs switch roles. The gate and source terminals are connected
together and driven by a photovoltaic diode stack in the optocoupler. Many photodiodes are connected in series so that the
sum of the individual photodiode voltages is greater than the
FET turn-on threshold voltage. The photovoltaic output is controlled by the illumination from the optically isolated LED. Field-
effect transistors can be destroyed by high-voltage line spikes
and must be protected. Protection is generally provided by a
varistor or transorb clamp placed across the FET terminals to
absorb the energy of high-voltage spikes by becoming low impedance past their voltage knee.

Much attention has been given to the bilateral FET solid-state
relay because of its near-ideal characteristics. The FET output,
being resistive, does not have the fixed voltage drop across its
terminals that is found in the other circuits. On-state resis-
tances of less than 1 Ω are common. Off-state resistance is typically hundreds of megohms, resulting in virtually no leakage
current. AC and DC are equally well switched. Bilateral FETs
are compatible with low-level TTL and CMOS logic gates. Switching occurs immediately, independent of line voltage or current
phase. Switching times are typically 100 μs or more, resulting

in elimination of electrical switching noise generation. RC snubbers are not required since there is no sensitivity to the dV/dt
effect. The bilateral FET switch is not capable of handling large
inrush currents. At currents just above the rated current, the
switch resistance rises quickly, causing a dramatic rise in voltage across the switch, producing a lot of heat within the FET.
The FET is somewhat more rugged than a bipolar transistor under these conditions, but can still easily become overheated and
destroyed if operated repeatedly or continuously beyond its current rating.

4.3 TRANSISTOR INTERFACES

Simple transistor circuits are the interface of choice in most low-
and medium-power DC applications. Transistor circuits have
proven their reliability in industry and are reasonably well understood due to their simplicity. Transistors are virtually ideal
DC switches having extremely high impedance in the off-state and
very low voltage drop in the on-state. Transistor interfaces are
generally the least expensive, fastest response, and present no
electrical danger to misplaced fingers.

4.3.1 Construction

A transistor interface is constructed of electronic components
having no moving parts. It is usually assembled as part of the
photoelectric circuit, but is often encapsulated in a plug-in module, as shown in Fig. 4.21. Most transistor output circuits are
driven directly from the photoelectric power supply and do not
isolate the field wiring circuits from the photoelectric, although
some models do offer plug-in optical isolation modules. The
transistor is a silicon device with three electrical terminals that
make electrical contact with three separate regions on the silicon. These regions are implanted with other specific "doping"
atoms to give them semiconducting properties. The result is a
device with a collector terminal output current that is controlled
by the base terminal input current. Two types of transistors,
NPN and PNP, are made, so that either positive or negative polarity current may be controlled. Their schematic symbols are
shown in Fig. 4.22. Collector current i_c is controlled by the
base current i_b, both of which flow through the emitter terminal.
Collector current is generally about 100 times as large as the base
current if it is not limited by a load in the collector circuit.

Figure 4.21 Plug-in transistor interface modules.

When the collector current is limited, the transistor is said to be saturated and will appear much like a closed relay contact with only a few tenths of a volt from collector to emitter. When there is no base current, the transistor turns off and no collector current flows. In the off-state, the entire load supply voltage is developed across the collector-to-emitter terminals just as it would on an open switch. Standard photoelectric transistor outputs are generally rated up to about 40 VDC blocking and usually

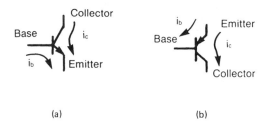

(a)   (b)

Figure 4.22 Schematic symbols of (a) NPN and (b) PNP transistors showing base and collector current paths.

not more than 250 mA conducting. Transistors are not regenera-
tive by nature as are triacs and SCRs and therefore cannot handle
large inrush surge currents beyond the continuous current rating.
A transistor circuit may function well in such a capacity by limiting inrush current. However, the response time of the load will
probably be increased, and repetitive cycling of the load may
overheat and destroy the transistor.

4.3.2 NPN Sink Transistors

The NPN transistor is most commonly used as a current-sinking
output device. That is, the NPN transistor is used to pull current through a load toward the lower-voltage side of the load
circuit. The four circuit configurations of Fig. 4.23 are quite
typical of DC photoelectric sensors.

The simplest of them is Fig. 4.23a, which is referred to as an
open-collector NPN. The collector of the transistor is uncom-
mitted and can be connected to loads with higher external supply voltages. For example, if the photoelectric control is operated from 12 V, the open-collector NPN transistor may drive a
relay powered by 24 V. In this case the user must take responsibility for protecting the transistor from the inductive kick voltage by connecting a clamp diode across the load terminals as
shown.

The circuit of Fig. 4.23b includes the clamp diode internally
for protection of the transistor. The assumption here is that
the load will operate from the same supply voltage as the photoelectric circuit, or from a lower-voltage supply. If this circuit
is used with the photoelectric control operating from 12 V and a
relay coil load operating from 24 V, once energized, the relay
may never turn off. Current will flow through the relay coil to
the collector, forward bias the clamp diode, and flow into the
photoelectric 12-V supply. Whenever the load supply voltage is
higher than the photoelectric supply voltage, this circuit will be
troublesome. One solution to this problem is the circuit of Fig.
4.23c, which uses a zener diode to clamp the collector voltage.
A zener diode blocks current flow until a specific voltage is
reached. At this point it becomes low in resistance and absorbs
the inductive kick energy itself rather than dumping the energy
into the power supply or allowing the transistor to be destroyed.
The zener diode will typically be rated to clamp at 36 V for 40-V
rated transistors. The zener diode solution is the best solution
for transistor protection since it is built in and does not limit application of the photoelectric control.

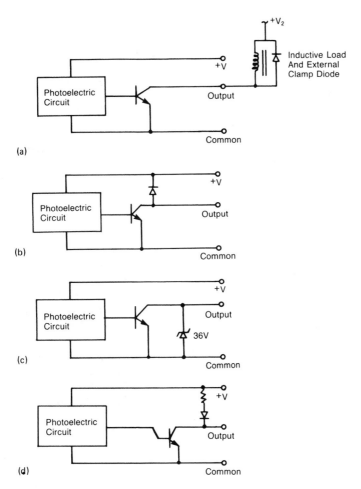

Figure 4.23 NPN transistor current-sinking circuits: (a) open-collector NPN showing inductive load with clamp diode; (b) NPN transistor with built-in clamp to supply; (c) NPN transistor with built-in zener diode clamp; (d) NPN transistor with pull-up resistor for logic interface.

Finally, the circuit of Fig. 4.23d offers a pull-up resistor. The pull-up resistor pulls the collector terminal high when the transistor is off so that it may present a change in output voltage to some external device. Most digital logic circuits require

a voltage change rather than just current sinking for their interface. Without the pull-up resistor, there will be no change in
the collector output voltage measured at the transistor switches
when measured with respect to common. The diode in series with
the resistor is used to prevent current from flowing through the
resistor path when the load circuit is driven from a higher voltage than the photoelectric supply. The pull-up resistor might
alternatively be connected to an internal 5-V supply to make the
output voltage swing directly compatible with 5-V logic.

4.3.3 PNP Source Transistors

The PNP transistor is most commonly used as a current-sourcing
output device. That is, the PNP transistor is used to push current through a load from the higher-voltage side of the load circuit. This output configuration is used when the load is already
connected to ground on one side, as is the case with many programmable controller input modules. The representative circuit
in Fig. 4.24 of a current-sourcing PNP transistor output is quite
similar to its NPN transistor cousin. All of the comments regarding transistor protection and pull-up resistors for the NPN apply
directly for the PNP and are connected appropriately for the PNP
polarity.

Perhaps the optimum DC output interface configuration is the
source/sink/TTL combination shown in Fig. 4.25. This circuit
has protected transistors and is capable of interfacing directly
with inductive or resistive loads connected to either supply voltage, common, or to 5-V logic. This circuit provides the greatest flexibility of application.

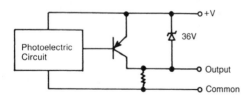

Figure 4.24 PNP transistor current-sourcing circuit.

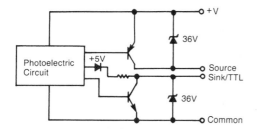

Figure 4.25 Application of flexible source/sink/TTL combination.

4.3.4 Optically Isolated Circuits

When running field wiring long distances through a factory, optical isolation is recommended to prevent electrical faults from propagating damage through a complex electrical control system. Optical isolation also eliminates ground loop currents, which can be nasty sources of electrical noise and erratic performance. Optically isolated DC circuits, represented in Fig. 4.26, are usually quite simple. Because they have no terminal in common with power or common, they may be connected as a current sink or current source. Generally, optocouplers are used directly as sinking or sourcing outputs for interfacing to low-current logic circuits, programmable controllers, electronic counters, other optical couplers, or external solid-state relays. They are typically capable of sinking 10 to 50 mA at about 0.4 V saturation for low currents in the on-state and withstanding 30 VDC in the off-state. When larger loads must be driven directly, the NPN phototransistor may be used to drive a second transistor in what is called the Darlington transistor configuration of Fig. 4.26b. This circuit may easily drive a few hundred milliamperes of current. Due to the Darlington configuration, on-state voltage will usually be nearly 1 V, making it unsuitable for most logic interfaces.

4.4 TWO-WIRE CONTROLS

Two-wire photoelectric controls are designed to provide a simple interface requiring only a two-wire connection to function, much like a simple mechanical switch. In fact, these designs are packaged almost exclusively in the same style housing as that of

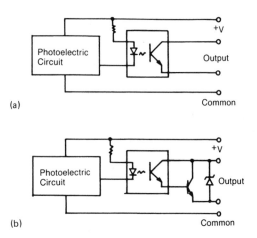

Figure 4.26 Optically isolated (a) low-current and (b) high-current Darlington NPN transistor interfaces.

standard industrial mechanical limit switches (Fig. 4.27). In effect, the result is a noncontact low-maintenance replacement for the mechanical limit switch workhorse of industry. It is said that there are two types of mechanical limit switches: brand new and broken. Two-wire switches offer simple unambiguous wiring, fewer wires to pull, familiarity in packaging and connection, direct programmable controller hookup, reliable long-life solid-state switching, and heavy-duty packaging. Some manufacturers additionally offer short-circuit protection, AC/DC operation, and multiple diagnostic LED indicators. With the cost of wiring estimated at between $1 and $10 per foot per wire, the higher cost of two-wire versus three- to five-wire photoelectrics is often easily justified. Two-wire switches have the disadvantage of reduced optical performance, high on-state voltage drop, and off-state leakage current.

4.4.1 Construction

Physically, virtually all two-wire optical limit switches are packaged in the familiar mechanical limit switch style body and are composed of three main components: the optical sensing head, power module half of the lower body, and wiring compartment part of the lower body, as shown in the disassembled view of

2-Wire Photoelectric Controls

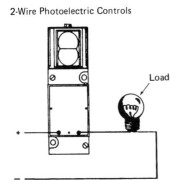

Load

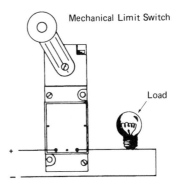

Mechanical Limit Switch

Load

+

–

Switch is wired in series with load.
Two wires are required.

+

–

When arm is mechanically
depressed, circuit is completed.

Figure 4.27 Two-wire photoelectric control with industrial mechanical limit switch and magnetic proximity switch cousins. (Courtesy of Opcon, Inc.)

Fig. 4.28. The power module has a relatively small compartment in which the power supply conditioning, output switch, and timing logic functions all reside. The two-wire connection to the wiring

Figure 4.28 Disassembled view of two-wire optical limit switch showing optical sensing head, power module, and wiring compartment.

compartment is made by two blades that are inserted into sockets in the wiring compartment as the halves are assembled. A small circuit board with many miniature and high-voltage electronic components is encapsulated within this lower power module. The encapsulation provides electrical isolation between components, the case, and the wiring. It also provides excellent mechanical support for the components in high-shock and high-vibration environments. Power dissipated by the switch in the form of heat is conducted directly to the case, which in turn is cooled by the surrounding air.

4.2.2 Electrical Performance

The ideal two-wire switch is the simple set of contacts shown in Fig. 4.29a. A two-wire electronic switch attempts to approximate this with a circuit equivalent to Fig. 4.29b. When the switch is open, the resistor provides a leak path for current around the switch to power the photoelectric electronics. When the switch is closed, the zener diode provides an on-state voltage high enough to power the electronics. The bridge rectifier converts AC load current to DC for switch operation.

   All two-wire electronic controls "steal" their operating power from the load circuit. This means that there must be some leakage current when the switch is off and some voltage drop when the switch is on. These switches must have leakage current small enough that they will not energize any rated load. Most operate with leakage current below the 1.7-mA limit required for a recognized off-state by most programmable controller AC input

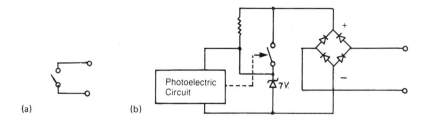

(a)   (b)

Figure 4.29 (a) The ideal two-wire switch is a simple set of contacts; (b) two-wire electronic switch approximation of the ideal.

modules. This means that very little power is available to operate the photoelectric circuitry, including the LED indicators, the
transmitter, the receiver, the logic module, and the switch circuit. This necessitates some compromises in performance. For
example, optical performance will be less than that of three- and
four-wire switches with similar optics, and high-intensity indicator lamps should not be expected, both because of limited available current to drive the transmitter and indicator LEDs.

The voltage drop required to operate the two-wire circuit
causes heat-generating power dissipation in the unit when the
load is energized. Typically, the voltage drop will be 7 to 9 V,
depending on the load and the manufacturer's design. Continuous load ratings therefore depend mostly on the ability of the
unit to operate hot. For this reason, manufacturers generally
derate their two-wire switches at higher ambient operating temperatures and load currents. Mounting the power module against
a metal object will help lower the power module operating temperature in heavy load circumstances by conducting away some
of the heat. In locations where ambient temperatures are high
or a great deal of radiant heat energy is present, special care
should be taken to ensure that power module case temperatures
do not exceed approximately 85°C (185°F).

Two-wire controls are wired in series with the load. Figure
4.30a shows that a load operating from 110 V will actually receive only 101 V when the switch is turned on because the

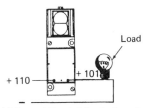

(a) Control "steals" 9 volts from load.
101 volts remains to drive load.

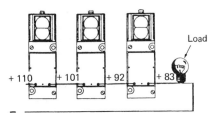

(b) Voltage drop accumulates when
units are wired in series.

Figure 4.30 (a) A two-wire photoelectric switch reduces the
available voltage to the load by the amount of the on-state voltage drop; (b) multiple switches each further reduce the voltage
available to the load. (Courtesy of Opcon, Inc.)

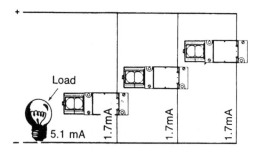

1.7mA leakage accumulates when
units are wired in parallel.

Figure 4.31 The leakage current flowing through the load is cumulative as two-wire switches are paralleled. (Courtesy of Opcon, Inc.)

switch will steal about 9 V from the load circuit for its own operation. The 9 V required for operation of the switch is cumulative
when more than one is connected in series with the load. As Fig.
4.30b shows, three switches in series would reduce the load voltage in the same circuit to 83 V. The number of switches that may
be wired in series depends on the minimum load voltage requirements and expected minimum line voltage under normal operating
circumstances, as some loads are more tolerant than others of reduction in operating voltage. The leakage current of multiple
two-wire switches connected in parallel with one another is cumulative, as shown in Fig. 4.31. The total leakage current that
will flow through the load when all switches are off will be the
sum of the leakage currents for each of the paralleled switches.
Three parallel switches each with 1.7 mA of leakage will produce
a total off-state leakage current of 5.1 mA. This is not a problem with high-current loads, such as contactors and lamps. However, the higher leakage current will likely cause most high-impedance loads, such as programmable controller input circuits,
to latch-on. Before connecting multiple two-wire switches in series or parallel, check with the manufacturer's specifications, since
not all two-wire switches have been designed to function properly
when connected in series or parallel with one another.
      The implementation of two-wire power module circuits varies
widely. Some are only capable of handling AC load circuits. These

have probably been implemented with SCR or triac output switches. These switches will handle easily the large inrush currents of contactors, solenoids, and lamps. They will also be slow in response due to zero-crossing switching delays. Other varieties use a power FET or transistor switch. These devices will switch rapidly, but generally will not have the same inrush current capacity. However, most specify some capability in this regard, but care must be taken to understand the application requirements and possible switch limitations.

4.5 ANALOG INTERFACES

Analog interface photoelectric sensors are designed to produce an analog output voltage or current proportional to received signal. They are capable of providing more information to a control system than simple presence/absence indication. Depending on the optical sensing configuration used, the analog output signal may be proportional to distance, material color, optical transmissivity, or lateral position of an object. They provide useful solutions to web loop control, cut-to-length control, edge guiding, measuring clarity of juices, and sizing of fruits and related applications where a "how much" type of signal is required. Few photoelectric manufacturers produce analog output photoelectric sensors. Comparatively, it is a special problem-solving sensor.

4.5.1 Circuit Adjustments

Most of an analog sensor's photoelectric circuitry is the same as that of the on/off versions. Figure 4.32 illustrates that following the receiver amplifier and peak detector, the usual threshold detector circuit is replaced with analog amplifier circuitry. To

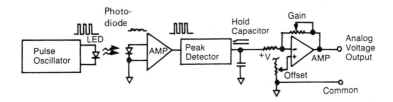

Figure 4.32 Block diagram of an analog voltage output photoelectric control.

interface well with other system controls, it is generally necessary to have offset and gain adjustments within the photoelectric
to tailor the analog voltage reference and sensitivity. For example, it may be required that a web edge be controlled to a certain position and that deviations from that position should drive
a correcting motor proportional to the deviation. The installation
would require that the web be adjusted to its desired position and
that the photoelectric offset adjustment be set to null the output
signal under these conditions. The gain is then set to provide
sufficient signal, on deviation from the desired position, to cause
the system to operate effectively and without instability. These
two adjustments are provided by all analog photoelectric sensors.
In many circumstances, however, two simple single-turn poten-
tiometer adjustments are not sufficient. The adjustments on some
designs interact with one another, and great difficulty may be
found in making required fine-tuning adjustments, especially after two cups of morning coffee. Analog photoelectric sensor models with noninteracting multiturn adjustments do exist and simplify installation. In addition, unless the circuit is internally
compensated, the output will drift with change in temperature
and age. Most circuits are not well temperature compensated,
while others have built-in circuits to compensate for LED variation with temperature and age. An LED will vary in light output
by a 2:1 ratio from -40 to +70°C and will probably decrease by
about 10% or more in intensity after one year of operation. These
effects can seriously affect the reliability or repeatability of an
analog operation application and are in addition to problems related to poor maintenance of lens cleanliness.

4.5.2 Output Circuits

There are two basic analog interface output circuits: voltage producing and current sourcing. Voltage-producing models output
an analog voltage directly from the final amplifier in Fig. 4.32.
The voltage range available varies considerably between models.
Some may output only between 0 and 5 V, while others are capable of a full -10 to +10 V. The output voltages are intended
for pilot duty only since they generally cannot drive more than
10 to 20 mA of current.

   The current-sourcing circuit of Fig. 4.33 is often chosen in
industrial environments for its superior noise immunity relative
to voltage-producing interfaces when field wiring is laid in conduit adjacent to other electrically noisy field wiring. Current-
sourcing outputs produce a source current to the load. In the

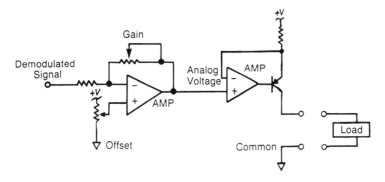

Figure 4.33 Current output interface, 4 to 20 mA.

circuit shown, the final amplifier regulates current through the
transistor by measuring the voltage generated across a resistor.
The standard current-sourcing interface for control circuits and
instrumentation is 4 to 20 mA. Broken circuits sourcing 0 mA
may be detected automatically. However, many simpler circuits
generally produce the full range from 0 to 20 mA.

These voltage-producing and current-sourcing interfaces are
not capable of directly driving any process control device (e.g.,
motors, heaters, or lamps) but are intended for interfacing with
programmable controllers, motor speed controllers, data acquisi-
tion modules, and other process control equipment that will actually provide the end power control signal.

5

Photoelectric Logic

'Contrariwise,' continued *Tweedledee, 'if it was so, it might be; and if it were so it would be; but as it isn't, it ain't. That's Logic.'*

Lewis Carroll
Through the Looking-Glass

In this chapter we describe in detail the timing, counting, output device, and photodetector logic used in photoelectric controls, how they operate, how their properties may be taken advantage of, and what pitfalls may be encountered in their application. Logic is a natural extension of simple sensors. Sensors detect an input event and control an output event. The output often requires different timing than that of the input event. Sometimes several inputs must jointly produce an output event. In other words, logic alters the raw sensor signal so that it is directly useful for controlling local action in response to sensed events. Today's highly automated factories generally have shifted logic functions from the sensor to a programmable logic controller or minicomputer program which supervises and controls entire machines and processes. Logic functions are often more easily maintained, better protected from tampering, and more cost-effective when implemented in the process controller program.

5.1 TIMING LOGIC

Timing logic plug-in modules are the most common reference to and usage of logic with photoelectric controls. Timing modules

Figure 5.1 Photoelectric timing logic modules.

conditionally filter the input signal or cause the output signal to be stretched, shortened, or displaced in time. Typical timing logic hardware shown in Fig. 5.1 is built into the photoelectric, plugged into the photoelectric as an option, or available as an entirely separate item. Timing adjustment is generally available over the range of 0.005 to 30.0s.

5.1.1 Time Delay Logic

Time-delay logic offers the ability to filter out or ignore short-duration events and act only on longer-duration events. There are three types of time delays: on delay, off delay, and on/off delay.

On-delay timing is shown in Fig. 5.2. On-delay timers delay the generation of an output signal by a preset time interval from the start of an input signal. Input signals shorter in duration than the preset delay interval will not generate an output signal. Output signals are terminated immediately upon termination of the input signal. On-delay timers are often used to detect conveyor jam conditions. Passing objects produce input signals shorter than the preset on delay interval. The input signal becomes con-tinuous when objects are jammed in the conveyor and soon re-sults in the generation of an output signal.

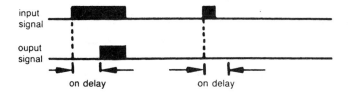

Figure 5.2 On delay.

Off-delay timing is shown in Fig. 5.3. Off-delay timers delay or hold the termination of an output signal by a preset time interval from the end of an input signal. Loss of the input signal for less than the preset off-delay hold interval will not cause undesired dropout of the output. The off delay can also be used as a simple output pulse extender when input signals are not of sufficient duration to be used directly in driving physical devices.

On/off delay incorporates both the on-delay and output-hold timing functions in one module, as shown in Fig. 5.4. Confusion often results when deciding whether an on delay or an off delay is required, since the detection of an object may result in either a high- or a low-going signal from the photoelectric to the logic module. For example, a fiber optic sensor may be set up to sense objects in proximity or in thru-beam. In proximity the pulsed light is received by the detector when an object is present. In thru-beam, the pulsed light is blocked from the detector when an object is present. The usual convention is to put it in terms of the signal as viewed by the photoelectric detector. Proximity sensors are most often used in a "light-operate" mode, while reflex and thru-beam sensors are used in a "dark-operate" mode. For example, delayed detection for a dark-operate thru-beam

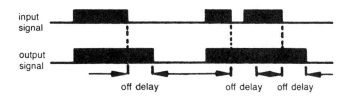

Figure 5.3 Off delay.

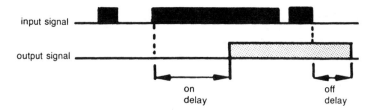

input signal

output signal

on
delay

off
delay

Figure 5.4 On/off delay.

means that the off-going photoelectric signal must be delayed or, in other words, the off-delay timer is required. To alleviate this problem, some photoelectrics offer light/dark operate selection prior to the timing function and/or following the timing function. Do not feel alone if you become confused trying to determine which end is up when working this problem.

Simple on/off delay timers designed with a single timing element for both on and off delay timing may suffer from a timing interaction problem unless special design precaution was taken to eliminate it. The problem is easily identified. Although timing may be accurate when the change input signal occurs over relatively long intervals, when the input signal switches state quickly after the end of a timing cycle, the following time cycle may be reduced by as much as 40%! This problem may become apparent in circumstances such as changing line speeds or object spacings on a conveyor. This design pitfall can be detected with a simple manual experiment. Both on and off delays are set to about 2 s; then the beam is alternately made and broken as described above. Fortunately, there are many timing circuits available that do not have this problem.

5.1.2 One-Shot Circuits

A one-shot circuit produces a single pulse of fixed length in response to a change in state of the input signal. The pulse length is fixed by the circuit components and is independent of the length of the input pulse. The one-shot can be set to trigger from either a light-to-dark or dark-to-light transition. As such, it is often referred to as an edge-triggered rather than a level-triggered circuit. Selection of edge triggering allows the circuit to produce a pulse at either the leading or trailing edge of a passing object. Since the output pulse length is fixed by the circuit,

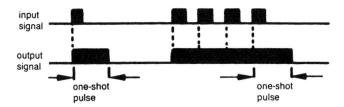

Figure 5.5 Retriggerable one-shot.

output pulses will not vary in length with changing line speeds. The value of a one-shot circuit is its ability to provide a fixed-length pulse output to solenoids and reject mechanisms, which often require consistent length control signals for optimum operation. For example, a blast of air to reject a defective bottle may not be successful if the pulse is too short and may reject two bottles if the pulse is too long.

The two basic types of one-shots are the retriggerable one-shot and the nonretriggerable one-shot. The retriggerable one-shot shown in Fig. 5.5 will restart the output pulse timing cycle and extend the pulse length if a trigger signal reoccurs during the output pulse. The output pulse can be extended indefinitely as long as the time interval between input triggers is shorter than the output pulse. Conversely, the nonretriggerable one-shot shown in Fig. 5.6 will not initiate a new output pulse timing cycle while an output pulse is currently being generated. This one-shot will produce consistent length pulses under all conditions. For example, nonretriggerable one-shots provide pulse-length consistency when undesirable spurious input signals such as multiple reflections from a glass bottle are received.

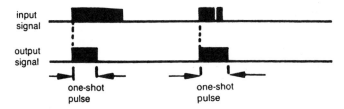

Figure 5.6 Nonretriggerable one-shot.

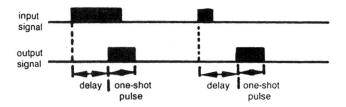

Figure 5.7 Delayed one-shot.

The delayed one-shot shown in Fig. 5.7 is actually composed of a pair of cascaded nonretriggerable one-shots. The first edge-triggered one-shot is used to produce an internal delayed trigger for the second one-shot, which produces the output pulse. A delayed one-shot is often used at inspection stations to provide a delayed rejection control signal for a downstream rejection mechanism. The delay interval is set to match the length of time it takes to transport the defective item from the inspection station to the rejection station. The one-shot output pulse is adjusted for optimum response of the rejection mechanism.

The on-delay one-shot shown in Fig. 5.8 requires an input signal to be present for a fixed period of time before a one-shot output pulse is generated. Input signals shorter than the on-delay time interval will not trigger the output one-shot. Once the one-shot has been triggered, the input signal may continue or be terminated with no effect on the output one-shot. This timing function is useful for rejection of stuck parts, which can be detected by their long duration.

Resettable one-shot timing logic, shown in Fig. 5.9, produces a one-shot output pulse limited in length by the one-shot timing

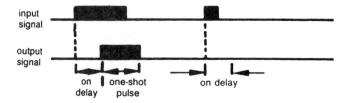

Figure 5.8 On-delay one-shot.

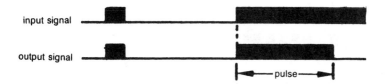

Figure 5.9 Resettable one-shot.

interval or the end of the input signal. This function is also
known as "time-limited on/off" logic or as an "interval timer."
This timing logic is used in energy or water conservation controls to limit the operation time of the device. Control of checkout conveyors at your favorite grocery store is a typical example.

5.1.3 Underspeed/Overspeed

Rate sensing logic provides the capability to monitor the rate or
speed of a machine, conveyor, or process by measuring the time
between sensed events. The time interval is measured between
either leading or trailing edges of each event. For example, a
photoelectric control can be installed so that the beam is made
or broken by a keyway in a shaft or a hole in a gear. Underspeed conditions are detected when the time between input signals becomes greater than the preset time interval, as shown in
Fig. 5.10. During system power-up, the output of underspeed
detecting logic output must be inhibited until the system has a
chance to reach its normal operating speed. When a conveyor
mechanism jams, underspeed logic can turn off the drive motor
before it is damaged from overload. Overspeed conditions are
detected when the time between input signals becomes less than

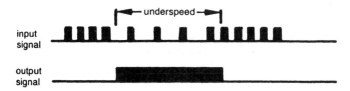

Figure 5.10 Underspeed detection.

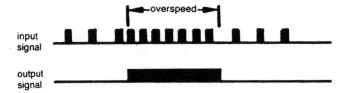

Figure 5.11 Overspeed detection.

the preset time interval, as shown in Fig. 5.11. Overspeed logic prevents potential runaway conditions by shutting the system down. True rate sensing logic produces stable outputs as a result of underspeed or overspeed conditions, as opposed to the erratic output that would be produced if a retriggerable one-shot was used for this function.

5.2 COUNTING LOGIC

Counting logic conditions the photoelectric output by the number of events sensed. Counting logic includes latches, preset counters, and totalizers, all of which are quantity oriented. Highly automated production facilities today generally include counting logic as part of the programmable logic controller or minicomputer program that supervises and controls entire machines and processes.

5.2.1 Latch

A latch is a single-event counter with the characteristics shown in Fig. 5.12. It is triggered by a single input signal and holds

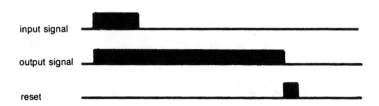

Figure 5.12 Latch.

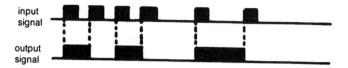

Figure 5.13 Divide-by-2 flip-flop.

the output latched in the same state until power is cycled or an external reset signal is received. A latch can be used to shut down a machine on detection of a malfunction, or in the control of asynchronous events. For example, a latch may be used to shut down a conveyor following the arrival of an object until the operator can physically attend to the object. When the conveyor is cleared, the operator presses the reset button, allowing the conveyor to bring the next object.

5.2.2 Preset Counter

A preset counter or divider produces a single pulse output after receiving a preset number of input pulses. The most common preset counters are the divide-by-2, -6, and -24. Some preset counter designs offer a limited selection of preset counts, while others offer selection of any number desired up to the maximum of the logic device. The divide-by-2 counter is often called a "flip-flop" and alternates between a high and a low output state following each input signal, as shown in Fig. 5.13. A typical application might be to alternate the position of a diverter on a conveyor to split the feed evenly into two packaging machines. Preset counters for higher counts produce a one-shot pulse after reaching the preset count and automatically restart the count sequence as shown in Fig. 5.14. Preset counters are widely used on packaging

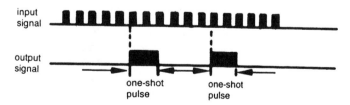

Figure 5.14 Divide-by-6 preset counter with one-shot output.

machines to control the filling of a package with the required number of items prior to sealing the package and indexing to the next package. For example, a six-pack of soft-drink cans may use a divide-by-6 to count out the number of cans per package and a divide-by-2 to order them in two columns.

5.2.3 Totalizing Counter

A totalizing counter does not interact with the output signal. It simply provides a total count of all input signals. Totalizers such as the one shown in Fig. 5.15 are often used to count the number of items produced in a particular production run or machine center. When another group of items are to be counted, the operator must reset the totalizer with an external reset signal prior to the start of the new batch. Usually, this is done using a wired electrical reset signal. However, for the unit in Fig. 5.15 it is accomplished by bringing a chain-attached magnet near the counter display window, where its magnetic field causes a reed switch inside the unit to close and reset the totalized count. Some totalizing counters are provided with a means to connect a battery for backup power to save the count in case of power failure.

5.2.4 Shift Register

A shift register is a series of electronic memory cells that are used to delay signals until a predetermined number of clock pulses have been received. For example, clock pulses may be generated by a conveyor sprocket and photoelectric switch (i.e., one pulse per quarter inch), which together with the shift register produce a delay that represents a known amount of movement rather than time. At each clock pulse the present state of the input signal is shifted into the first memory cell and at the same time shifts its old data to the second cell. Similarly, each additional memory cell receives its data from the cell immediately preceding it, in bucket brigade fashion. The last cell produces the actual output, which is a delayed version of the input signal. Thus the conveyor can run at fluctuating speeds without affecting the location of the object when the output signal occurs. The delay interval is set by selecting the length of the shift register and the clocking signal. For example, a shift register with 100 cells and clocked once every quarter inch will produce an output after a delay of 25 in. Figure 5.16 shows a shift register delayed signal requiring 11 clock pulses to pass the input signal through the shift register to the output. A principal virtue of the shift

Figure 5.15 Totalizing counter with magnetic reset. (Courtesy of Opcon, Inc.)

register is that it will remember multiple items in transit on the same conveyor. For example, if a conveyor of bottles spaced at 2 in. has a rejection station 12 in. from the inspection station and two consecutive bottles are sensed as having missing caps, a delayed one-shot will not reject the second bottle. The second input trigger occurs before the delayed one-shot has finished dealing with the first input and will be ignored by the one-shot. A delayed one-shot also will not produce the rejection signal at the proper time if line speed is varied. A shift register will perform both operations correctly. However, it is necessary for the

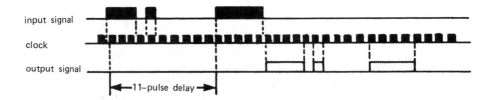

input signal

clock

output signal

11-pulse delay

Figure 5.16 Shift register output delayed by 11 pulses.

clocking rate of the shift register to be quite a bit faster than the
maximum rate that input signals occur to ensure sufficient resolution in the sampling of the input signal. A shift register will not
reliably detect an input pulse if it is shorter than the clock pulse
interval, and it will not produce an output that is shorter than a
clock pulse interval because of the interval sampling nature of
the serial memory shift register circuitry.

5.3 OUTPUT DEVICE LOGIC

Output signals from more than one sensor often must be combined
to generate the required system control output. Quite often it is
possible to produce the required output through direct hard-wired
combination of the individual sensor outputs. Output device interconnection can provide great utility in control simplification and
cost reduction for both relay and solid-state output devices. Although combinational logic often is easily accomplished through
connecting the output devices of multiple photoelectric controls
to drive a single load, poor training and incomplete information
generally have limited its use.

An important fact to keep in mind when dealing with output
device logic is that a TRUE or ON-state output may represent
either object presence or absence. Often the user has the choice,
through selection, of light-operate or dark-operate outputs. Although an output may be desired only when two objects are present, incorrect selection of light/dark operation may produce an
output only when both objects are absent. Embarrassing errors
may be prevented by thinking this matter through thoroughly
beforehand.

There are, in addition, possible side effects from some of these
interconnections of which one should be aware. These include:
excessive voltage drop in series-connected switches, excessive

Table 5.1 Logical OR Output as a Function of
All Possible Combinations from Three Inputs[a]

Input			Output
A	B	C	Q
0	0	0	0
0	0	1	1
0	1	0	1
0	1	1	1
1	0	0	1
1	0	1	1
1	1	0	1
1	1	1	1

[a] 0 = FALSE/OFF; 1 = TRUE/ON.

leakage current in parallel-connected switches, and potential interaction between multiple two-wire switches.

5.3.1 Parallel OR

The term OR, in combinational logic of two or more devices, defines the resultant output as being TRUE or ON if one OR or more
of the inputs is TRUE. Table 5.1 shows OR combinational logic
for three inputs, any or all of which, when TRUE, produce a
TRUE output.

The OR output logic function is produced by connecting output switches in parallel as shown in Fig. 5.17. Here the switches
have been schematically shown as normally open relay contacts.
Clearly, any one switch OR any combination of switches is capable of switching power to the load. The parallel OR configuration is useful when an expanded or split sensing region is required for control of a single actuator. It can also be used to
reduce the number of input modules required for programmable
controller interfacing. Optically isolated solid-state switches,
including isolated transistors, FETs, and triacs, have both terminals available for connection and may be interconnected in the
same parallel manner as the relay contacts shown in Fig. 5.17.

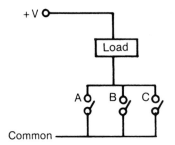

Figure 5.17 Parallel OR-connected relay contacts.

The circuit-equivalent symbols for these devices are shown in Fig. 5.18. Electrical polarity for the DC transistor switch must, of course, be connected appropriately.

Transistor outputs used on 10–30 VDC sensors generally have NPN sinking or PNP sourcing, or both transistor types. NPN sinking transistor outputs already have their emitter terminals connected to circuit common while PNP sourcing transistors already have their emitter terminals connected to the positive DC voltage supply. As a consequence, NPN sinking transistors may only switch in parallel to circuit common, and PNP sourcing transistors may only switch in parallel to the DC supply voltage, as shown in Fig. 5.19. NPN transistor outputs that

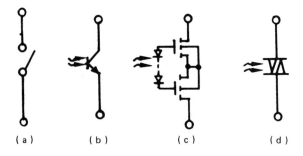

Figure 5.18 Circuit-equivalent switch symbols: (a) relay; (b) isolated NPN transistor; (c) isolated bilateral FET; (d) isolated triac.

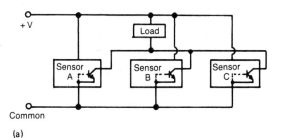

(a)

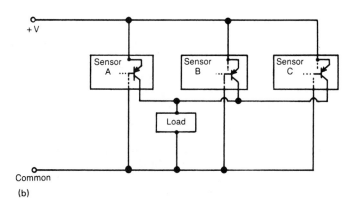

(b)

Figure 5.19 (a) Paralleled open-collector NPN sinking transistors;
(b) paralleled open-collector PNP sourcing transistors.

have an internal pull-up resistor are intended to be compatible
with voltage-sensing logic-level circuits. The internal pull-up
resistor raises the voltage at the output terminal when the transistor turns OFF. When these outputs are parallel connected as
shown in Fig. 5.20, the resultant output voltage to the logic-level
circuitry is low whenever any of the transistors are ON, and high
only when they are all OFF. This is referred to as inverted logic
since the standard representation of ON or TRUE in logic-level
circuitry is a high-level voltage, while OFF or FALSE is represented by low voltage. Note that output-device pull-up resistors that raise the input voltage to a logic gate above the logic
circuitry supply voltage may either damage the logic circuit or
cause it to malfunction. CMOS logic gate inputs are the most
sensitive to this problem.

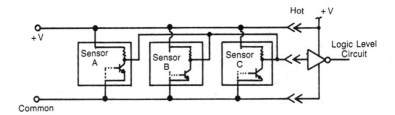

Figure 5.20 NPN transistors with pull-up resistors interface directly with logic-level circuitry.

Two-, three-, and four-wire AC switches may be parallel connected to switch power from either AC hot or neutral to the load. Because they are AC devices, it does not matter how they are powered from the AC line. However, it is safest to switch AC hot to the load. Most two-wire AC/DC switches may be operated in parallel (check the manufacturer's specifications to be sure). Figure 5.21 shows that the same sensors can be parallel OR connected for switching AC hot or common. Three- and four-wire AC switches that use a triac as the switching element generally use an RC snubber to eliminate switching dV/dt latch-up when driving inductive loads. However, the RC snubber causes a leakage current path around the switch. Two-wire switches use OFF-state leakage current to operate the switch's photoelectric sensor electronics. When these switches with leakage current are operated in parallel, as shown in Fig. 5.22, their leakage currents sum into one larger leakage current. In the figure the leakage currents from the individual switches sum to 5.1 mA. Leakage current is not a problem for high current loads such as contactors and light bulbs, but can be problematic for the high-impedance input circuits used on electronic control equipment. Most programmable controller AC input circuits require less than 2 mA of input current to detect an OFF-state reliably. In our example, total OFF-state leakage current is enough to cause these programmable controller input circuits to erroneously detect an ON state, probably resulting in improper operation. The leakage current, however, may be shunted around the high-impedance input by placing a resistor in parallel with the input terminals. For 115 VAC connections, a resistor value of 22 kΩ at 1 W will successfully shunt about 1.7 mA around the programmable controller input. To shunt three times the current, use three of these resistors or a single 7.5-kΩ 2-W resistor.

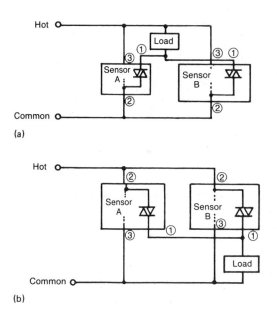

Figure 5.21 Three-wire AC sensors may be connected in parallel to switch either (a) AC hot or (b) common to the load.

5.3.2 Series AND

The term AND, in combinational logic, defines the resultant output as being TRUE or ON only when all the inputs are TRUE. Table 5.2 shows this result for three inputs. A TRUE output is produced only when all inputs are simultaneously TRUE. In other words, only when input A AND input B AND input C are all TRUE will the output be TRUE. The logical AND configuration is used primarily in gating functions. For example, a sensor with a one-shot output may be positioned to detect the leading edge of a bottle. A second sensor may be aligned to test for label presence by using a thru-beam sensor. By logically ANDing their outputs, the resultant output produces a reject signal only when a bottle is actually present AND when a label is missing.

The AND output logic function is produced by connecting the output switches in series as shown in Fig. 5.23. Here the switches have been shown schematically as normally open relay contacts. Clearly all of them must be ON in order to switch power to the

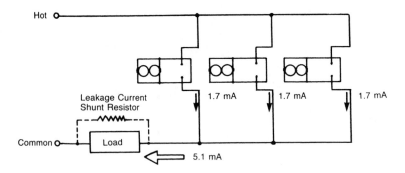

Figure 5.22 Sum of leakage currents from parallel-connected AC switches. A parallel resistor shunting a high-impedance input circuit may be required.

load. Optically isolated solid-state switches, including isolated transistors, FETs, and triacs, have both terminals available for connection and may be interconnected in the same series manner as the relay contacts shown in Fig. 5.23. The circuit equivalent

Table 5.2 Logical AND Output as a Function of All Possible Combinations from Three Inputs[a]

	Input		Output
A	B	C	Q
0	0	0	0
0	0	1	0
0	1	0	0
0	1	1	0
1	0	0	0
1	0	1	0
1	1	0	0
1	1	1	1

[a]0 = FALSE/OFF; 1 = TRUE/ON.

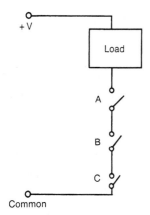

Figure 5.23 Series AND-connected relay contacts.

symbols for these devices are shown in Fig. 5.18. Electrical polarity for the DC transistor switch must, of course, be connected
appropriately.

As indicated in the preceding section, dc-operated sensors
with NPN sinking transistor outputs have their emitter terminal
already connected to circuit common, while PNP sourcing transistors have their emitter terminal already connected to the positive DC voltage supply. As a consequence, NPN transistor outputs may not be stacked together nor may PNP transistor outputs be stacked together. Both the collector and emitter terminals of the transistor must be available to accomplish this, as
is true with optically isolated transistor output modules. However, Fig. 5.24 indicates how a NPN sinking output may be combined with a PNP sourcing output to produce the logical AND
function. One transistor completes the load circuit to common,
while the other completes the load circuit to supply voltage. This
configuration is not useful for most DC programmable controllers
and general logic input circuits since they also have a terminal
precommitted to common or supply.

Three-wire AC switches have the same basic limitation as the
DC sinking and sourcing outputs because they also have a precommitted output terminal. They may, however, be used to produce the logical AND output by similar methods, as shown in
Fig. 5.25.

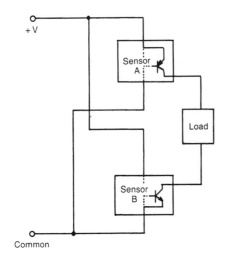

Figure 5.24 NPN sinking and PNP sourcing outputs in logical AND configuration.

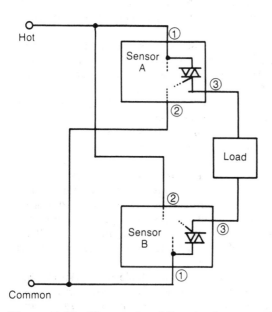

Figure 5.25 Three-wire AC outputs connected in logical AND configuration.

Most two-wire AC/DC switches may be stacked in series in the same manner as may multiple sets of relay contacts (check the manufacturer's specifications to be sure). Two-wire switches have an ON-state voltage drop that varies between manufacturers from 6 to 11 V. When the switch is ON, it requires this voltage drop to operate the sensor electronics. When these switches are put in series, voltage that would otherwise be delivered to the load is decreased by the ON-state voltage for each switch in series. For example, if a single switch has an ON-state voltage drop of 9 V, then as shown in Fig. 5.26, three such switches in series operating from 115 VAC will deliver only 88 VAC to the load. That is 27 VAC or a 24% loss in voltage and a 42% loss in delivered power!

5.4 PHOTODETECTOR LOGIC

Photodetector logic is the connection of one or more photodetectors to a single demodulation amplifier in order to perform multipoint sensing with a single output. These configurations can

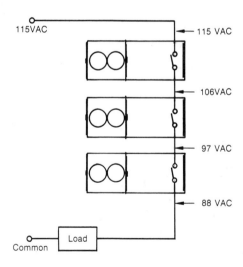

Figure 5.26 Voltage drop across multiple two-wire switches is cumulative in the series AND configuration. Here 9 V is lost across each switch.

only be offered in systems with remotely connected optical heads. Not all demodulation amplifiers manufactured are designed to allow this capability, so be careful to check with the manufacturer before creating a monster that smokes.

5.4.1 Parallel OR

Photodetectors connected in parallel produce the logical OR function. All parallel-connected detectors contribute to the signal input to the amplifier, as shown in Fig. 5.27. If any one or more delivers signals above the threshold detection level, the demodulation circuitry produces an output signal. In other words, any of the detectors can independently or together provide enough signal to cause detection by the demodulator, resulting in a change of output state. In proximity, an object is detected by the presence of light. In reflex and thru-beam, an object is detected by the absence of light. Although with regard to the photodetector, it is always an OR proposition, with regard to the application it may be either OR or AND. For proximity, an object present in position 1 OR 2 OR ... will cause detection. For reflex and thru-beam, all detectors must be blocked for the output to switch. This requires object presence in position 1 AND 2 AND .... These relationships are shown in Table 5.3. As customary, the outputs in the table are shown with proximity as light-operate and reflex and thru-beam as dark-operate, since an output is generally expected as a result of object presence.

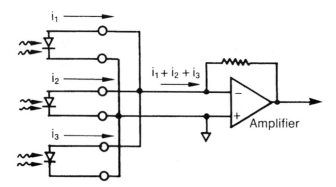

Figure 5.27 Parallel-connected photodiodes or phototransistors produce the logical OR of their received signals.

Table 5.3 Logical OR Photodetector Signal Responses[a]

| Proximity (light-operate) | | | | | Reflex and thru-beam (dark-operate) | | | | |
| Sensor[b] | | | Ampli- | | Sensor[b] | | | Ampli- | |
3	2	1	fier[c]	Output	3	2	1	fier[c]	Output
A	A	A	D	Off	A	A	A	L	Off
A	A	P	L	On	A	A	P	L	Off
A	P	A	L	On	A	P	A	L	Off
A	P	P	L	On	A	P	P	L	Off
P	A	A	L	On	P	A	A	L	Off
P	A	P	L	On	P	A	P	L	Off
P	P	A	L	On	P	P	A	L	Off
P	P	P	L	On	P	P	P	D	On

[a]Logical OR photodetector signals effectively produce the logical OR response in proximity and the logical AND response in reflex and thru-beam. The entries indicate the condition at the sensor, the amplifier input, and the output.
[b]Object status: A, absent; P, present.
[c]Detector status: D, dark; L, light.

   Remote parallel connections with either phototransistors or photodiodes can be used with any manufacturer's amplifier, due to the nature of the connection required. In both cases, the leads of one detector head are connected directly to the leads of the same color from other detector heads. This type of parallel connection is possible because it does not interfere with bias voltage or current conditions imposed on the photodetector by the amplifier. However, there are two important effects for which caution must be exercised: ambient light immunity, and threshold shift or latch-up. Both of these are related to the summing nature of the parallel-connected photodetectors. Photodetectors generate current proportional to the amount of light falling on them. Amplifiers have a limited tolerance for DC current generated as a result of ambient light. This is the ambient light immunity specification, usually given in terms of footcandles (fc). Sunlight at high noon on a clear summer day when reflected off a diffuse 90% white reflector can produce nearly 10,000 fc. An amplifier rated for 10,000 fc of light immunity with one detector will only be able to tolerate 2000 fc when five detectors are connected in parallel, since the dc current generated in both cases is the same. Similarly, because the detector photocurrents are summed, three detectors, each producing only one-third of the threshold photocurrent signal, will together reach threshold and cause a change in output state. Clearly, to prevent this problem, it is important that each of the parallel-connected detector heads be used in high-contrast applications. Thru-beam sensing applications have little problem here since beam-blocking objects are generally quite opaque. Thru-beam setups, however, can be troubled by multipath reflections around the object from nearby surfaces or translucent blocking objects. Parallel reflex sensors, each with its respective reflector blocked by an object, may each receive just enough proximity signal from the object's surface that the collective signal will never get below threshold. Similarly, proximity sensors may each receive enough signal from background objects to alter sensitivity or cause sensor latch-up. In most every case, the problem is solved by reducing the detector sensitivity, which lowers the "dark signal" below threshold.

5.4.2 Sum/Difference

Photodiodes, with some restrictions, may be connected either in parallel or in inverse parallel, as shown in Figs. 5.27 and 5.28, to produce the sum or difference of their respective signals. The inverse parallel connection is limited in use to amplifiers that do not apply a bias voltage to the photodiode. Photodiodes are often reverse biased in order to increase circuit immunity to diffuse sunlight illumination. A bias voltage applied to an inverse parallel-connected photodiode would drive the inverse diode into conduction, causing virtually all signal current to be diverted through the photodiode rather than to the amplifier. Phototransistors require a bias voltage in order to operate. Phototransistors will not operate at zero bias, nor will they operate when the bias polarity is reversed. Thus phototransistors cannot be used in inverse parallel configurations under any circumstances. The analog sum or difference configurations find their greatest use in servo controls, process controls, fixed-range sensing, and low-contrast detection. The sum configuration is most useful in extending the field of view of a detector by using additional

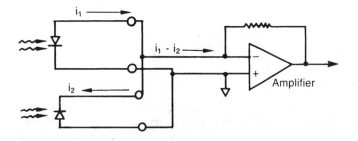

Figure 5.28 Inverse parallel connections of photodiodes can be used to create analog difference signals useful with low-contrast amplifiers.

detector heads to view additional areas of interest. The difference configuration has tremendous utility. A second detector in inverse parallel, like that of Fig. 5.29, can perform the function of a reference by subtracting off signal common to both detectors, leaving only the difference. This technique provides excellent compensation for effects common to both detector heads, such as LED aging, dust buildup, temperature variation, and process variation. In this way, only real differences from the reference detector are sensed. In fixed-range sensing applications, the inverse parallel detector is set up as in Fig. 5.30, so that the inverse parallel detector field of view causes the net signal to turn sharply at a fixed range. This method of range sensing is extremely stable. The range is effectively set by the geometry defined by the physical setup of the optical heads and is quite independent of target reflectivity and optical path degradation. Careful setup of a differential detector system can yield impressive repeatability results with process control, servo control, low-contrast, and fixed-range sensing applications.

5.4.3 Series AND

Series AND connections for photodetectors, as shown in Fig. 5.31, may seem possible on the surface, but are not a functional configuration in practice. In almost all applications, some ambient light is present that will cause dc-generated photocurrent. The current that will flow through the two series-connected photodetectors will be limited by the smaller of the currents that each detector will try to develop. In total absence from ambient light,

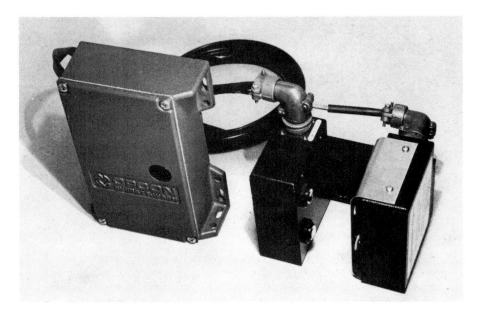

Figure 5.29 An inverse parallel dual detector sensor head compensates for process and environmental variations to produce a very stable high-sensitivity low-contrast detector. (Courtesy of Opcon, Inc.)

the series AND-connected photodiodes will work fine. The presence of ambient light allows a detector receiving no pulsed LED illumination to pass the pulsed signal generated by the second photodetector. The logical AND function is disrupted by the dc photocurrent and is not a recommended configuration!

5.4.4 Fiber Optic Logic

A single fiber optic photoelectric control can sense multiple points through the use of special fiber optic cables that split the transmitted light into multiple cable tips and then recombine the light from multiple receiving tips back into a single receiving fiber. There are many forms that can easily be imagined for unique applications, many of which will require a special fiber optic tip design. Special fiber tips can be be manufactured easily and economically with a few weeks lead time and are capable of solving otherwise difficult sensing problems.

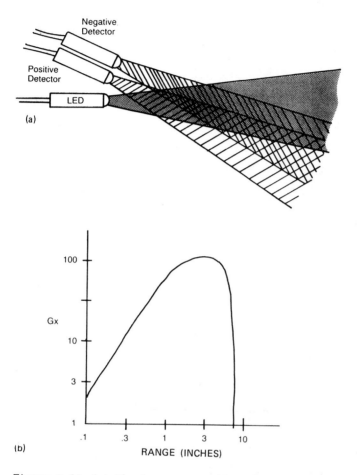

Figure 5.30 (a) Fixed-range sensing using a second inverse parallel detector produces (b) a very sharp cutoff excess gain sensor.

The simplest of fiber optic logic requires only a different use of the standard bifurcated cable. Two bifurcated cables may be inserted, as shown in Fig. 5.32, so that both source and detector cable tips are bifurcated. This results in two light-transmitting tips and two light-receiving tips that can be used as a pair of thru-beam fiber optic sensors. The photoelectric output will change states only when the light path between fiber tip pair 1 AND fiber tip pair 2 are both interrupted by objects. Special

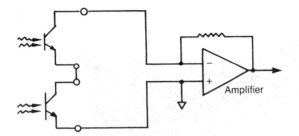

Figure 5.31 Series AND connection of photodiodes and phototransistors does not work in practice because dc current generated by ambient light nullifies the logical AND function for ac signals.

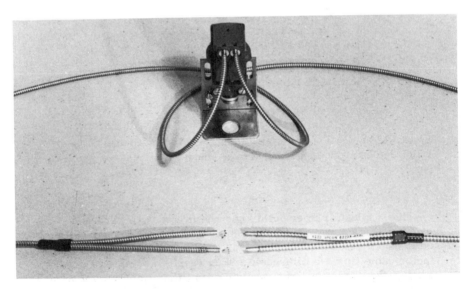

Figure 5.32 Two standard bifurcated cables may be used to produce a dark-operated logical AND thru-beam sensing pair.

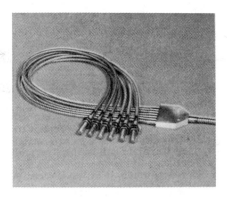

Figure 5.33 A single fiber bundle cable may be split into many branches to optically provide a multiple-input dark-operated logical AND thru-beam system. (Courtesy of Banner Engineering Corporation.)

fiber tips, such as those in Figs. 5.33 and 5.34, may be special ordered to expand this principle manyfold. When the original fiber bundle is split only once, the sensing range is reduced by a factor of 2. The range reduction for any number of splits may

Figure 5.34 A special split-tip configuration designed for integrated-circuit pin inspection. (Courtesy of Banner Engineering Corporation.)

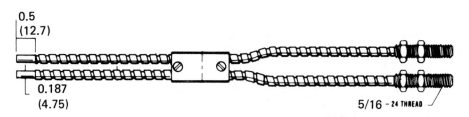

0.5
(12.7)

0.187
(4.75)

5/16 - 24 THREAD

Figure 5.35 A double bifurcated cable optically provides multiple-input OR proximity sensing. (Courtesy of Dolan-Jenner Industries, Inc.)

be estimated simply by dividing the single fiber sensing range by the number of splits. For example, a 12-in. range fiber optic sensor split six ways would be reduced to a 2-in. sensing range each.

Multiple proximity sense points are also possible with a single photoelectric. The double bifurcated cable shown in Fig. 5.35 is constructed by splitting each of the source and detector bundles into two separate bundles and combining each of the half source bundles with a half detector bundle to result in a pair of proximity sensor tips. The output of the photoelectric will change state if an object is present at either tip 1 OR tip 2. This configuration may also be manufactured with more than two sensing tips. The sensing range is reduced proportionally by the number of splits, as it was with the split thru-beam fiber optic tips discussed previously.

5.4.5 Multiple LEDs

Multiple LEDs used with multiple detectors to do photoelement logic should always be connected in series, regardless of the photodetector connection scheme. Figure 5.36a indicates that when LEDs are connected in parallel, the drive current is shared unequally between them. Even if the individual LEDs did share the current equally, the brightness of each would be reduced proportionally. There are three major reasons for unequal sharing of current. First, there are variations from LED to LED in the amount of current that will flow through them at a fixed forward bias voltage. Second, differences in the lengths of the cables would reduce the current flow in those with longer cables. Third, the current that flows at a given LED forward bias voltage is quite temperature

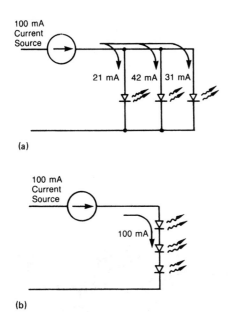

(a)

(b)

Figure 5.36 (a) LEDs in parallel divide the current unequally between them; (b) LEDs in series are all driven by the same current.

sensitive and would pose significant repeatability problems were there any temperature gradients between LED heads. For these reasons, photoelectric circuits are designed to drive LEDs with a constant current source rather than a voltage source. Connecting the LEDs in series, as shown in Fig. 5.36b, causes all LEDs to receive the same current drive regardless of their individual differences. The number of LEDs that may be driven in series is limited to the voltage compliance of the current source circuitry that drives the LEDs, as shown in Fig. 5.37. For example, a LED current source running from a 6-V supply may only be able to drive four LEDs in series. If each LED requires about 1.5 V, four LEDs require 6 V to drive the current through them. It would require 7.5 V to turn on five LEDs. None would turn on if our drive circuitry had only 6 V of compliance. Check the product data sheet or check with the manufacturer's application engineers to verify the feasibility of your multi-LED hookup if there is any question regarding your application.

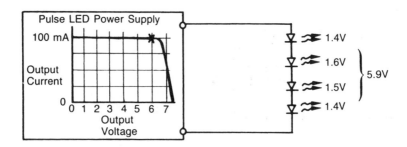

Figure 5.37 The number of LEDs that can be driven in series depends on the voltage compliance of the LED drive circuitry. Too many LEDs will completely prevent current from flowing.

Figure 5.38 Series and parallel connections of photoelements are aided by piggyback terminals designed for this purpose.

Parallel and series field connection of photoelements is greatly simplified with the piggyback connectors shown in Fig. 5.38 or simple terminal blocks. Other splices should be soldered for long-term reliability. Source and detector splices and connections should never be jumbled together. Separation of source and detector cabling prevents intercable crosstalk that can cause sensor latch-up.

6
Testing Standards

It is much less expensive to prevent errors than to rework, scrap, or service them.

Philip Crosby
Quality Is Free

In this chapter we describe various standards of performance against which photoelectric sensor and controls are tested. The value to you in becoming aware of these test standards is that of being able to identify readily photoelectrics that will perform well in your application. These tests relate to product safety, compatibility with the environment, and performance. Some of the tests are derived from recognized standards organizations, others are suggested standards generated specifically for photoelectric sensors and controls as the result of industry experience. Relatively simple methods for verification testing in the field will be described in lieu of the more time-consuming and equipment-intensive methods expected of photoelectric manufacturers. A list of relevant test standards appears in Appendix B.

6.1 OPTICAL STANDARDS

Not having been addressed directly by testing and standards agencies, optical testing standards for photoelectric sensors and controls have been left for definition to the individual manufacturers. The effect on performance of sunlight, artificial lighting, target size, and target reflectance can be nil or minor, or can totally prevent suitable operation. In this section we address these concerns and offer standards for their test.

6.1.1 Sunlight Immunity

Sunlight immunity is the ability of a photoelectric sensor to operate with direct sunlight falling on objects in the field of view of its detector. For example, a photoelectric control counting passing polar bears on a bright sunny winter day in Alaska should be tolerant of the extremely intense diffusely reflected light from the bear and the snow. Directly pointing the photoelectric at the sun, however, will not only overload and blind the receiver circuitry, but is likely to damage the detector or other nearby receiver circuit components as they fry in the intense heat generated by the focused solar light energy. My eyes do not function well either when looking directly into the sun, but have no trouble resolving a twig on a snow drift. The illumination produced by bright sunlight on a 100% white diffusely reflecting surface is 8900 fc on June 21 at noon at 30° north latitude according to Fig. 6.1. This is the basis for the common specification of 10,000 fc of sunlight immunity.

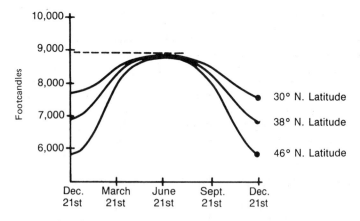

Figure 6.1 Solar illumination on a surface perpendicular to the solar rays at 12:00 noon. [From *IES Lighting Handbook*, 5th ed. (New York: Illuminating Engineering Society, 1978).]

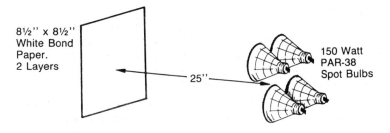

8½'' x 8½''
White Bond
Paper.
2 Layers

25''

150 Watt
PAR-38
Spot Bulbs

Figure 6.2 Simulating 10,000-fc illumination with standard incandescent spotlights.

Sunlight is often not conveniently available when we want to perform this simple sunlight immunity test. I prefer to eat lunch at noon, and in winter we have rain while other parts of the country are freezing in the sunshine. As an alternative, Fig. 6.2 shows four 150-W PAR-38 spot bulbs (commonly available at hardware stores) mounted 25 in. from a white diffuse target, as measured from the bulb surface, that will nicely approximate 10,000 fc of solar illumination. The lamps should be aimed to achieve the best uniformity of intensity across the target. The lamps must be operated at rated voltage to achieve the correct illumination. A 10% difference in line voltage will produce about a 30% difference in illumination. Calibration can easily be achieved with a photodiode and an ammeter by comparing actual readings in the sun with those from the lamps. Two layers of standard white bond paper 8.5 in. on a side make a nearly ideal diffuse 90% white reflecting target. For testing purposes, lower levels of illumination may more accurately simulate the actual environment. Indoor illumination through a window rarely exceeds 5000 fc. Vertical surfaces will not receive greater than 2000 fc of illumination from the sun due to the angle of incidence and/or absorption through the atmosphere at low solar altitudes. To simulate less solar illumination, one or more bulbs may be turned off.

Given a 10,000-fc illuminated target, we next require a test procedure for proximity, reflex, and thru-beam sensors and a criterion by which to judge the results. Total sunlight immunity is the ideal objective; however, this implies that absolutely no detectable effect will be observed—ideal but not generally realistic. A suggested standard that is practical and tolerable for user and manufacturer alike may be stated as:

Sunlight Immunity:

> The sensing range of a photoelectric sensor shall not vary by
> more than 10% and the excess gain shall not vary by more than
> 20% from that of low ambient lighting levels when the field of
> view of the detector includes a white diffuse reflecting surface
> illuminated by a 10,000-fc light source.

Testing for a change in the sensing range is not very difficult.
However, testing for a change in excess gain will require inter-
polation from the excess gain versus range performance graph
supplied by the manufacturer, except in the case of thru-beam
and long-range proximity models, where the relationship may be
approximated as: The percent change in excess gain equals twice
the percentage change in range.

Proximity models are simply tested using the illuminated white
diffuse paper reflector as the target. The maximum sensing range
is measured twice, with the lights on and lights off, by moving
the photoelectric sensor toward the white paper target until it
is first detected. The percent change in range is given by the
formula

$$\text{\% range change} = 100\ \frac{(\text{range off}) - (\text{range on})}{\text{range off}} \qquad (6.1)$$

Thru-beam sensors are tested in a proximity fashion by aiming
both the source and detector at the white paper target. The
range is measured by using the detector as the moving sensor
while the source remains at a fixed distance. The source is
mounted near the target and projects a relatively small spot of
pulsed light onto the center of the paper target. Although the
thru-beam sensor is actually being used here as a sort of proximity sensor, the method works. Reflex sensors are tested by
placing a 1-in. plastic corner cube reflector on the center of the
paper target and again measuring the range by the same procedure. The reflex sensor field of view will overlap the edges of
the retroreflector at maximum range, thus taking the bright paper into part of its field of view. The reflex sensor should be
at least 20° off-axis from the lamps to be sure the reflex sensor
does not receive a retroreflection from the lamps. The plastic
corner cube retroreflector style must be used when measuring
the performance of polarized reflex units.

6.1.2 Fluorescent Light Immunity

Fluorescent light immunity is a measure of the detection circuit's
ability to reject pulsating light signals produced by other light

sources in the local environment. This test is particularly important since many factories are illuminated with fluorescent, mercury
vapor, or other lamps that produce significant bursts of light at
120 cycles per second and at its higher-frequency harmonics as a
result of being powered from AC line voltage. Even though the detector circuit may be synchronized to the pulsed LED circuit, other
very strong signals may still be capable of causing detection. The
suggested test standard may be simply stated as:

Fluorescent Light Immunity:
    A photoelectric sensor shall not demonstrate an unstable flick-
    ering detection state at any range when pointed directly at the
    surface in the center of an operating 40 W daylight or cool
    white fluorescent lamp.

This simple test lends confidence that the sensing process will not
be affected by the presence of strong illumination in the detector's
field of view from other artificial lighting sources, particularly
when near the threshold of detection.

6.1.3 Excess Gain Specifications

There are three significant optical performance specifications for
photoelectric sensors: excess gain versus range, field of view,
and sensing zone. The excess gain curves described in Chapters
2 and 3 are generated using specific test targets. Excess gain
curves are plotted versus range on log-log scale graph paper.
Field measurements of excess gain are difficult to make. Manufacturers generally acquire these data through access to internal
analog signals or through use of calibrated attenuated reflectivity targets.

Reflex Excess Gain:
    Excess gain versus range graphs for reflex photoelectric sen-
    sors shall be produced on log-log scale graph paper using data
    acquired from 3-in.-diameter and 1-in.-diameter plastic corner
    cube retroreflectors with relative reflectivity of 5000, and with
    a 1-in. square retroreflective tape target specified by the
    manufacturer.

    The industry standard target for the performance specification
of a reflex sensor has been a 3-in.-diameter molded plastic corner
cube retroreflector. Miniature photoelectric controls intended for
short-range operation should also specify performance with a 1-in.-
diameter molded plastic corner cube retroreflector and a 1-in.-
square retroreflective tape target. These targets have significantly

different performance characteristics, are more widely used with
miniature photoelectrics, and could make or break the success of
an application. The reflectivity of a plastic corner cube reflector
is typically close to 10,000 times as bright as a diffuse white reflector. However, distribution measurements of hundreds of retroreflectors show that if we are to guarantee performance with
98% confidence, targets with reflectivities as low as 5000 must be
used for test purposes. Performance of retroreflective tapes
varies widely among the many types available. Information on
reflectivity and environmental durability of tapes used should be
specified by the sensor manufacturer so that alternative tapes
may be used as necessary with predictable results through tape
performance comparison. Examples of excess gain specifications
are presented in Fig. 6.3.

Proximity Excess Gain:
    Excess gain versus range graphs for proximity photoelectric
    sensors shall be produced on log-log scale graph paper using
    data acquired from 90% diffusely reflecting targets of size 10,
    3.3, 1.0, 0.33, and 0.10 in. square, as appropriate. Specifying
    additional specific target data is recommended for special pur-
    pose proximity sensors.

The industry standard target used to specify performance of
a proximity sensor has been the 90% white diffusely reflecting
Kodak test card R-27, catalog number 1527795. The size of the
test card is rarely specified by manufacturers, but may generally be assumed to be at least as large as the effective beam diameter at all measurement ranges in order to show the best possible performance. However, photoelectric sensors are more often used to detect small objects than large boxes on a conveyor,
particularly special purpose proximity sensors such as focused
proximity, short-range proximity, and wide-angle proximity. Specification of these should additionally include targets more likely
to be encountered. This may include a smaller target for focused
proximity, or a glass jar for wide-angle proximity. Additional information provided by the sensor manufacturer will improve the
application ease of these products. An example graph for a long-
range proximity sensor is presented in Fig. 6.4.

Thru-beam Excess Gain:
    Excess gain versus range graphs for thru-beam photoelectric
    sensors shall be produced on log-log scale graph paper.

The excess gain for thru-beam sensors is not dependent on
target reflectivity or target size. A thru-beam sensor pair that

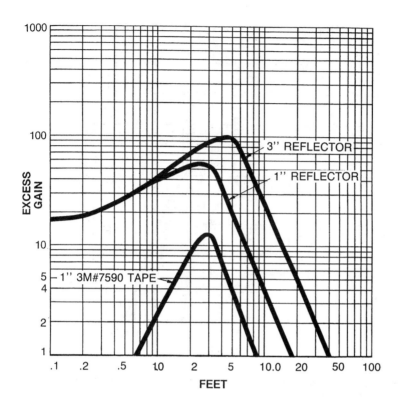

Figure 6.3 Example of excess gain performance specification for a reflex sensor.

is operating with sufficient margin will always require nearly complete blockage of the effective beam for object detection. The thru-beam curve may simply be shown as a -2 slope line on the log-log scale graph paper which bends toward horizontal at short distances that are approximately the size of the source and detector lens diameters. An example graph for a 30-ft thru-beam sensor with 0.30-in.-diameter lenses is presented in Fig. 6.5.

6.1.4 Field of View and Sensing Zone Specifications

While excess gain curves can answer the question of how well a target can be detected, the field of view and sensing zone

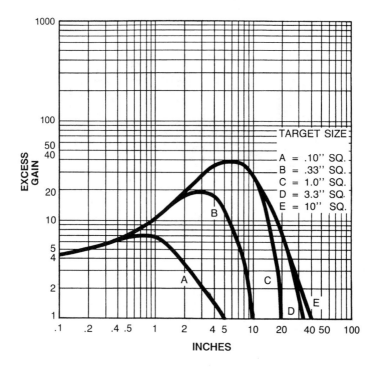

Figure 6.4 Example of excess gain performance specification for a long-range proximity sensor.

specification complement it by specifying where the target can be detected. The field of view specification tells where a sensor's light is going and where its detector is looking. For reflex and thru-beam sensors, this information tells us the spot size projected or detected at a given range, and the required pointing accuracy for alignment. For proximity sensors, the field of view and the sensing zone are nearly identical. In previous chapters, the term "effective beam" was used to describe a simple sensing zone in which the field of view of the source and detector overlap and objects may be detected. Here we will be more specific: The term "sensing zone" will be used to describe a more detailed relationship which includes effects of excess gain and target size.

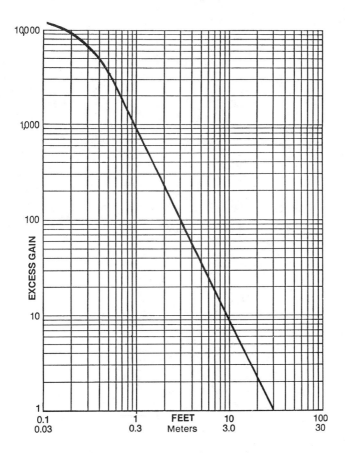

Figure 6.5 Example of excess gain performance specification for a 30-ft thru-beam sensor with 0.30-in.-diameter lenses.

Field of View:

    Field of view is an angular measurement in the far field of the detector sensitivity, or LED radiation pattern, where the angle specifies the full angular width at half of maximum sensitivity or intensity. Field of view is specified as an angle or as a spot size at a given range.

The field of view is a statement on how the source light energy is distributed and in what region the detector is sensitive.

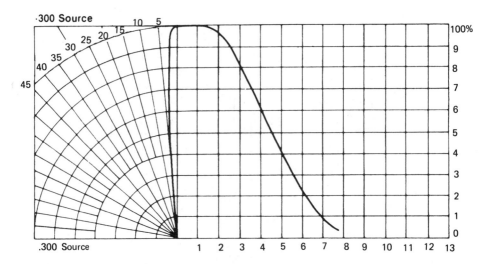

Figure 6.6 Relative sensitivity versus angle for a lensed detector. (Courtesy of Opcon, Inc.)

Generally, the field of view can be expressed easily in terms of a cone with a specified apex angle, or as a specified spot diameter at a fixed range. Given that the shape is conical, either means of specification provides the same information. The sensitivity of a detector or intensity of a source over its field of view may be mapped as shown in Fig. 6.6. The standard means of measuring the width of such a curve is to determine the full width at half of maximum sensitivity. As a general rule, the field of view of sensors with larger lenses is less than that of their smaller counterparts. In fact, if some basic assumptions are made, one finds that the field of view is inversely proportional to the lens diameter of the sensor. The same amount of light will be emitted by a small- and a large-lensed sensor; however, the large-lensed sensor concentrates this light energy into a smaller angular field of view, enabling it to increase its operating range at the expense of making alignment more difficult. In addition to giving information on alignment ease, the field-of-view specification provides information on how big a hole is required when it is necessary to sense through a bulkhead or wall. When calculating such a hole size, the idea to remember is that (except for the focused proximity) the spot size starts out at the size of the lens diameter and diverges into a cone as specified by the angular field of view.

Sensing Zone:

The sensing zone is specified by plotting, for each range, the lateral offset of the target's leading edge when detection is first achieved as it moves toward the beam axis. The range is measured from sensor to target for proximity sensors, source to detector for thru-beam sensors, and sensor to retroreflector for reflex sensors. The target shall move directly adjacent to the retroreflector surface for reflex units. Proximity units shall use the same targets as those used to specify excess gain.

The sensing zone specification is a descriptive tool borrowed from inductive proximity sensors. It is a pair of symmetrical curves that specify the sensing zone in terms of a specific target that is moved toward the axis of the beam from the side until it is detected. The position of the leading edge of the target and the target distance are measured as shown in Fig. 6.7 and the resultant data are graphed as shown in Fig. 6.8. The two lines cross where the leading edge of the target has penetrated halfway into the beam. If the target is at least as large as the sensor spot size at this range, this is also the range where the excess gain curve will cross 2.0. Beyond this range, the target must penetrate more and more into the beam until, at the range where excess gain is unity, it is fully centered. At close ranges where the excess gain is high, only a small portion of the effective beam need be penetrated to achieve detection. These effects combine to create a set of curves that generally has the appearance of a fish outline. Figure 6.8 shows the sensing zone for a

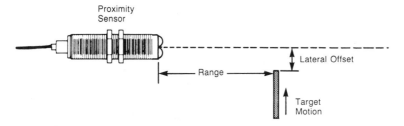

Figure 6.7 Specification of photoelectric proximity sensing zone.

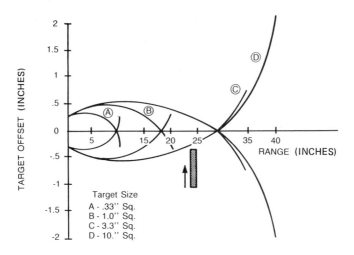

Figure 6.8 Sensing zone for a proximity sensor with a 2-in. field of view at a 20-in. range and unity excess gain at 40 in.

proximity sensor with a 2.0-in. field of view at a 20-in. range using square white targets 10, 3.3, 1.0, and 0.33 in. on a side.

With some modification in procedure, the same technique can be used to graph the sensing zone for thru-beam and reflex sensors. For each of these it is the effective beam that is interrupted by an intruding object. What the sensing zone curves indicate is how much of the effective beam must be interrupted for detection. For thru-beam sensors, Fig. 6.9 shows that the range is measured as the distance between the source and the detector. The location of the interrupting object between them is largely irrelevant. Figure 6.10 shows a sensing zone specification for a typical 400-ft range thru-beam photoelectric sensor. Here again, the two lines cross when the excess gain of the system is 2.0. Over most of the range, nearly the entire beam must be blocked to achieve detection.

For reflex sensors, the situation is a little more complicated since the effective beam usually varies in diameter between the sensor lens and the retroreflector. To make this technique useful, we must assume that the beam-interrupting object will be in close proximity to the retroreflector, as shown in Fig. 6.11. With this assumption made, the sensing zone may be graphed with the same procedure used for the thru-beam sensor. Figure 6.12

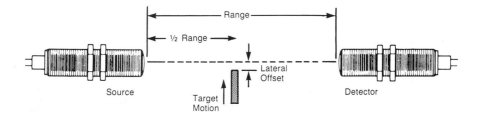

Figure 6.9 The sensing zone specification for thru-beam sensors specifies the range as the distance between the optical heads.

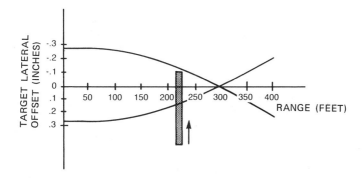

Figure 6.10 Sensing zone for a 400-ft range thru-beam sensor.

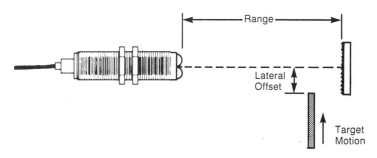

Figure 6.11 The sensing zone specification for reflex sensors assumes that the beam-blocking object is adjacent to the retroreflector.

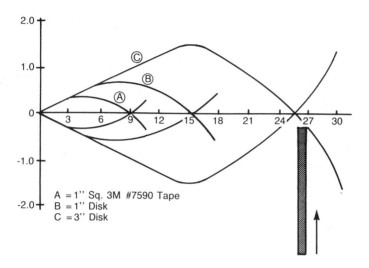

Figure 6.12 Sensing zone for a 30 ft range reflex sensor.

shows the sensing zone specification for a reflex sensor used with
3 in.- and 1 in.-diameter corner cube retroreflector disks, and a
1-in. square of beaded retroreflector tape.

6.2 ELECTRICAL STANDARDS

Electrical test standards are available for testing noise immunity,
performance, and safety. In this section we address the important electrical tests, and suggest general standards for them and
simple tests for their field verification.

6.2.1 Conducted Noise Immunity

Conducted electrical noise is produced primarily by sources such
as switching regulators for motors or power supplies, and showering arcs produced when relay contacts open inductive load circuits. High-frequency noise riding on the supply voltage leads
can potentially disrupt the highly sensitive photoelectric demodulation circuitry, resulting in false triggering.

Showering Arc Immunity:

> A photoelectric sensor shall not demonstrate an unstable flickering state of detection at any sensing range when subjected to the showering arc electrical noise characteristic of industrial environments when tested to NEMA ICS 1-109.

NEMA test standard ICS 1-109 is a severe test, but an excellent method for ensuring showering arc immunity. A calibrated arc is generated across a pair of open contacts powered by a neon sign transformer and other components. The showering arc current is run through 100 ft of cable in parallel with the power or load leads for the sensor to simulate the electrical noise coupling that occurs in industrial wiring trays. Figure 6.13 shows showering arc test equipment being used to test a proximity photoelectric control as it is cycled on and off by a rotating target.

Figure 6.13 NEMA ICS 1-109 showering arc test fixture.

A less sophisticated alternative field verification test for showering arc noise immunity is the chattering relay test. A relay is wired with its normally closed contacts in series with the relay coil so that it will oscillate, chattering on and off when power is applied. The chattering action produces many rapid asynchronous showering arc bursts. The chattering relay is connected at the end of a 15-ft length of household extension cord along with the sensor under test, as shown in Fig. 6.14. The extension cord serves to stabilize somewhat the impedance of the power line. Relays of different sizes and construction produce slightly different forms of electrical noise. An ordinary DPDT 10-A relay rated for the same voltage supplied to the sensor under test is adequate.

Supply Ripple Voltage Immunity:

A photoelectric sensor shall not demonstrate an unstable flickering state of detection at any sensing range when subjected to power supply ripple voltage of 200 mV peak to peak over the frequencies of 10 kHz to 1 MHz characteristic of industrial switching power supplies.

Switching power supplies have fundamental frequencies that may vary from 10 kHz to as high as 1 MHz and may put out as

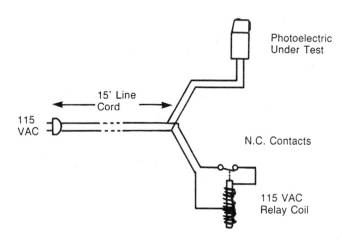

Figure 6.14 Chattering relay test wiring diagram.

much as 200 mV peak to peak at these frequencies. The showering arc and chattering relay tests do not directly test compatibility with industrial switching power supplies. They lend significant confidence for overall conducted noise immunity of a sensor,
but they will not test for particular problem frequencies generally found near the harmonic frequencies of the photoelectric
LED pulse frequency. The recommended test uses a signal generator transformer coupled in series with the sensor supply voltage, as shown in Fig. 6.15, as a means of generating ripple on
the power supply voltage. A 4:1 or 10:1 transformer designed
for this frequency range is recommended. The signal generator
is adjusted to produce 200 mV peak to peak across its secondary
terminals and is swept across the frequency range 10 kHz to 1
MHz.

6.2.2 Radio-Frequency Noise Immunity

Radio-Frequency Noise Immunity:
   A photoelectric sensor shall be neither turned on nor turned
   off by the presence of 5-W radio-frequency transmissions between 25 and 1000 MHz when the radiating antenna is 1 ft from
   the sensor or its cable in any orientation.

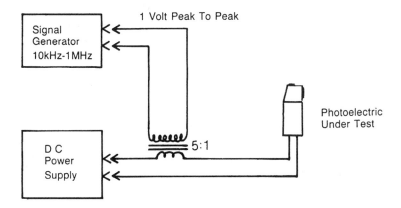

Figure 6.15 Transformer-coupled injection of ripple voltage into
the dc supply voltage.

Signals radiated from walkie-talkies and other communication devices in industrial plants can unintentionally disrupt the operation of nearby equipment. The amplifier circuitry in sensors of all types is prone to inadvertent pickup and detection of electromagnetic radiation unless preventive steps were taken in the sensor's design. The FCC has set the maximum radiated output power limits for the industrial class walkie-talkies at 5 W. At a 1-ft range, 5 W corresponds to a 50-V/m electric field. This requirement is more stringent than that for equipment around most military equipment and space launch vehicles! The practical test recommended is to subject the sensor to radio-frequency emissions by keying the transmitter of a walkie-talkie to frequencies likely to be used in the presence of the sensors at a 1-ft range from both the body and power cable. The photoelectric should neither turn on nor turn off in the presence of these radio-frequency emissions.

6.2.3 Transient Voltage Spike Immunity

Transient Voltage Spike Immunity:
   A photoelectric control shall not be permanently damaged when subjected to a 6-kV voltage spike superimposed on its AC power or load circuit as specified by ANSI/IEEE C62.41-1980.

AC power lines are electrically very dirty. Voltage spikes are superimposed on power-line voltage by the opening of large inductive load circuits, power grid switching by the power company, and lightning strikes to power lines. Although 6000-V power line spikes are rare, as indicated by Fig. 6.16, they can easily destroy unprotected electronic components. The test waveform shown in Fig. 6.17 has an open-circuit crest voltage of 6000 V and a short-circuit current of 200 A. Although these huge transients last only a few millionths of a second, they would certainly destroy unprotected circuits in their path. This test cannot be simulated by simple test equipment. The voltages produced are dangerous enough to cause permanent bodily shutdown of inattentive or poorly trained operators with their fingers in the wrong place at the right time. This test is best left to the manufacturer.

6.2.4 Output Ratings

The output rating of a photoelectric control is usually specified in terms of its off-state blocking voltage and leakage current,

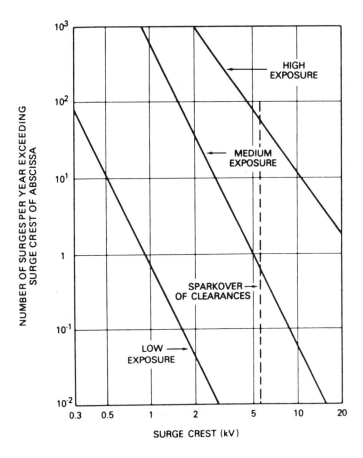

Figure 6.16 Rate of surge occurrences versus voltage level at unprotected locations. In some locations, sparkover of clearances may limit the overvoltages. (Courtesy of Institute of Electrical and Electronics Engineers, Inc. from ANSI/IEEE Standard C62.41-1980.)

and its on-state current capacity and voltage drop. However, a few other parameters must also be specified for AC circuits to ensure the success of an application. These include repetitive inrush surge current, short-circuit and overload protection, and load power factor compatibility.

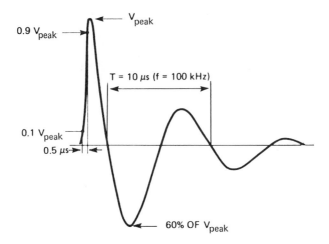

Figure 6.17 A 0.5-μs 1000-kHz ring wave, 6-kV open circuit, 200-A short circuit used to simulate lightning strikes. (Courtesy of Institute of Electrical and Electronics Engineers, Inc. from ANSI/ IEEE Standard C62.41-1980.)

Inrush Current:

>   Inrush current capacity of a photoelectric control, if tolerable, shall be specified at more than one current value, each with the associated repeat interval and duty cycle limitations required. Repeat intervals of 0.1, 1.0, and 10 s are suggested.

Inrush current is drawn by light bulbs, motors, solenoids, electrical contactors, and other electrical devices when they are first energized. Some types of solid-state relays easily handle surge currents 10 times their maximum continuous current rating, while others are barely capable of 50% over the continuous rating. Photoelectric control manufacturers often provide a nonrepetitive surge current rating; however, real applications almost always require the output device to operate more than just once during its rated life. The problem is that large inrush currents instantaneously cause a fixed temperature rise in the power switching component. When switched repeatedly, heat is generated more quickly than it can be dissipated. Excessive temperature eventually alters or permanently damages the switching component, rendering it nonfunctional. The key is in keeping the internal temperature of the power switching component below its rated safe

operating temperature. Short-interval repetitive output cycling
is quite common and can safely be accomplished with inrush currents in excess of the continuous current rating when the switching component conducts no current for part of that interval. However, short-interval repetitive output cycling with inrush currents near the rated maximum nonrepetitive inrush current will
probably lead to a very short functional life of the output switching component.

Short-Circuit Protection:
   Short-circuit-protected solid-state outputs on photoelectric
   controls shall withstand 5 minutes of continuous short circuit
   to rated supply voltage on the output terminals and shall withstand 5 minutes of 50% duty cycle switching into a short circuit at half-second intervals without permanent damage to the
   unit under test.

There are two basic methods of short-circuit protection of
solid-state output circuitry: current limiting and current foldback. Current-limiting circuitry measures the output current and
starts reducing the drive signal to the output switching device as
the output current passes a fixed current threshold. This holds
the output current at a prescribed maximum set by the internal
circuit components as long as the shorted output condition persists.
This circuit type can produce a lot of heat in a short period of
time and may literally burn out if the shorted condition persists.
However, there is value even in momentary short-circuit protection
in cases where wires are only momentarily bumped or shorted.

The technique used most often on high-voltage switching outputs is the current foldback method. This circuit detects current
levels rising past an internally set threshold and totally shuts
down the output switching device until the shorted condition is
corrected. The suggested test covers both the continuous short
that may arise on initial installation power-up, and repetitive
short conditions that may arise when a load fault occurs in a working installation. Both tests are required to prove the design ro-
bustness for both fault conditions.

Output Overload Fail Safe:
   An overloaded or short-circuited output, regardless of rating,
   shall not fail with dangerous outbursts of any kind that may
   be construed as a safety hazard to personnel in contact with
   the unit under test.

Most photoelectric controls purchased today are not protected
electronically from damage due to overloads or short circuits.

However, these controls also encounter load fault conditions. Consequently, the photoelectric control must be designed to at least fail in a safe manner when taken beyond specified operating limits. This is a destructive test and may not be compatible with your budget. Product certification by independent safety testing organizations such as Underwriters Laboratories and the Canadian Standards Association are clear indicators of compliance.

Load Power Factor:
   Photoelectric controls with AC outputs shall be tested for compatibility with load power factors of 1.0, 0.35 leading, and 0.35 lagging for load currents of 100%, 33%, and 10% of rated current.

The load power factor is an indirect measure of how much the current through the load leads or lags the phase of the AC supply voltage. When load current leads or lags the load voltage, the actual power delivered to the load is less than the voltage-ampere value by the cosine of the difference in phase angle. For example, if the load is inductive and the current lags the voltage by 60°, then only 50% of the voltage-ampere value will be actual delivered power, making the power factor equal to 0.50 lagging. Our concern here is not so much that the power factor may be less, but that not all AC switching devices are able to switch loads reliably with leading or lagging current phases. In particular, triac and SCR switches require special circuit design precautions to prevent latch-up of the switching element with low-power-factor loads. Contactors, solenoids, and other coiled wire loads are sources of lagging power factors. The capacitance between long cables and capacitance of programmable controller input circuits are sources of leading power factors. Capacitive or inductive loads with small resistive components produce very low power factors and are generally the most troublesome to switch reliably.

6.2.5 No False Pulse

When power is first applied to any electronic control, the circuitry takes a finite amount of time to reach a steady-state condition. In some applications it is extremely critical to have an output energized only when true detection occurs. For example, counters, ejection mechanisms, and other devices that produce physical action may disrupt operations or cause physical damage to goods or personnel if activated erroneously at power-up. Precautions must be taken in the circuit design to disable the output until the circuit has settled and the beam status (light or dark) has been acquired. Power-down is generally not considered a problem since any false pulse will occur after the power is no longer available to activate the load. However, in some cases the load circuit may operate from a different power source than the photoelectric. In this case, power-down false pulse protection may also be required.

No False Pulse:
    A photoelectric sensor or control shall not respond falsely during supply power application. The output shall remain inhibited until circuit transients have stabilized and beam status has been determined by the detection circuitry per NEMA ICS 2-229.

An issue yet to be resolved by industry is the desired power-up condition for timing modules. These devices produce outputs in the future based on signals generated in the past. Since the past is not known, there exists ambiguity. For example, an on-delay timer may first start its delay cycle on power-up if the beam is made. But since the object detected may have been there before power-up, the output timing to a printing or ejection device will probably be wrong. No matter what assumption is made, it will sometimes be wrong. Experimentation is generally required to determine the best solution for the application. It is possible to alter the initialization sequence of timing logic by swapping light/dark states on the input and output of the logic module and changing the logic function to accommodate the change (i.e., on delay to off delay). The best compromise available is to make sure that time-delay and one-shot modules receive no change in state during power-up. In other words, the normal sensor detection state should produce an inactive or off-state output to the timing module's input.

6.2.6 Safety

Safety in the workplace should always be a top priority. Safety certification of a product by safety testing or standards organizations is only the start. Certification indicates only that the manufacturer was conscientious with the design. It does not imply that the product can be safely misused. Safety is everyone's responsibility all the time.
    There are two levels of safety. The first level is the safety of the design under normal operating conditions. Are people reasonably protected from electrical shock, burns, and dangerous moving objects? The second level is the safety of the design under

failure conditions. Will failure cause an explosion or fire? Will
operational failure endanger human safety? For electrical safety,
Underwriters Laboratories, Inc., UL-508, Industrial Control
Equipment Standard for Safety, along with the Canadian Standards Association, CSA C22.2 No. 04-M1982, General Requirements—Canadian Electrical Code, Part II, are the two major safety
standards recognized in North America for certification of industrial control equipment. Machine guards for human safety are
regulated by standards set by the Occupational Safety and Health
Administration, OSHA 1910.217 and American National Standards,
ANSI B11.1-1982. The Robotic Industries Association has developed ANSI/RIA R15.06-1986, American National Standard for Industrial Robots and Robot Systems—Safety Requirements, which
specifies requirements for and of safeguarding devices. Numerous other organizations have published safety standards for their
special areas of concern, many of them being a related technical
society or association.

6.3 MECHANICAL/ENVIRONMENTAL STANDARDS

Suitability of a sensor for the environment of application will determine the service life of the sensor in that environment. A
sensor may be subjected to dust, water sprays, oil sprays, humidity, temperature extremes, mechanical shock, and vibration.
Understanding the conditions at the installation site is a prereq-
uisite to selection of a sensor so that the longest possible service
life may be ensured.

6.3.1 Enclosures

The recognized standard for industrial enclosures for electrical
equipment is NEMA Standards Publication No. 250-1979. NEMA
defines an enclosure as "a surrounding case constructed to provide a degree of protection to personnel against incidental contact with enclosed equipment, and to provide a degree of protection to the enclosed equipment against specified environmental
conditions." The key feature of each NEMA type enclosure is
listed in Table 6.1 for reference. The enclosure types of greatest importance to photoelectric sensors and controls are highlighted. Tables 6.2, 6.3, and 6.4 provide, in greater detail,
summary comparisons for specific applications of these enclosures.
Many of the higher-numbered enclosure ratings are inclusive of
other, lower-numbered ratings. Most enclosures meeting one
rating often automatically meet other, less stringent ratings.

Table 6.1 Primary Enclosure Characteristics of NEMA Standard
250-1979 and Equivalents in DIN Standard 40050[a]

Standards	Protection Level
IP 20, *NEMA 1*	*Fingers*
IP 22, NEMA 2	Falling dirt and water
IP 53, NEMA 3	Windblown dust, rain, and sleet
NEMA 3R	Falling rain and sleet
NEMA 3S	Windblown dust, rain, sleet, mechanisms operate iced over
IP 65, *NEMA 4*	*Hosedown*
NEMA 4X	Hosedown and corrosion
NEMA 5	Dust and falling dirt
IP 67, *NEMA 6*	*Temporary submersion*
IP 68, NEMA 6P	Occasional prolonged submersion and corrosion
NEMA 7	Indoor hazardous class I, groups A, B, C, or D
NEMA 8	Outdoor hazardous class I, groups A, B, C, or D
NEMA 9	Indoor hazardous class II, groups E, F, or G
NEMA 10	Mine safety
NEMA 11	Oil seepage, and corrosion
NEMA 12	Oil seepage
NEMA 12K	Oil seepage, has knockouts
NEMA 13	*Oil sprays*

[a]Types of greatest interest are highlighted.

One enclosure rating not covered by the NEMA types is a high-
pressure, high-temperature washdown enclosure required to meet
the requirements of the food and beverage industry. A survey of
the industry indicates that there are no current standards set by

Table 6.2 Comparison of Specific Applications of Enclosures for Indoor Nonhazardous Locations

Provides a degree of protection against the following environmental conditions	Type of enclosure										
	1[a]	2[a]	4	4X	5	6	6P	11	12	12K	13
Incidental contact with the enclosed equipment	×	×	×	×	×	×	×	×	×	×	×
Falling dirt	×	×	×	×	×	×	×	×	×	×	×
Falling liquids and light splashing		×	×	×	×	×	×	×	×	×	×
Dust, lint, fibers, and flyings[b]			×	×	×	×	×		×	×	×
Hosedown and splashing water			×	×		×	×				
Oil and coolant seepage						•			×	×	×
Oil or coolant spraying and splashing											×
Corrosive agents				×			×	×			
Occasional temporary submersion						×	×				
Occasional prolonged submersion							×				

[a]These enclosures may be ventilated. However, type I may not provide protection against small particles of falling dirt when ventilation is provided in the enclosure top. Consult the manufacturer.
[b]Fibers and flyings are nonhazardous materials and are not considered the class II type ignitable fibers or combustible flyings. See footnote b of Table 6.4.
Source: Reproduced by permission of the National Electrical Manufacturers Association from NEMA Standards Publication No. 250, *Enclosures for Electrical Equipment (1000 Volts or Less)*, copyright 1985 by NEMA.

Table 6.3 Comparison of Specific Applications of Enclosures for Outdoor Nonhazardous Locations

Provides a degree of protection against the following environmental conditions	Type of enclosure						
	3	3R[a]	3S	4	4X	6	6P
Incidental contact with the enclosed equipment	X	X	X	X	X	X	X
Rain, snow, and sleet[b]	X	X	X	X	X	X	X
Sleet[c]			X				
Windblown dust	X		X	X	X	X	X
Hosedown				X	X	X	X
Corrosive agents					X		X
Occasional temporary submersion						X	X
Occasional prolonged submersion							X

[a] These enclosures may be ventilated.

[b] External operating mechanisms are not required to be operable when the enclosure is ice covered.

[c] External operating mechanisms are operable when the enclosure is ice covered.

Source: Reproduced by permission of the National Electrical Manufacturers Association from NEMA Standards Publication No. 250, *Enclosures for Electrical Equipment (1000 Volts or Less)*, copyright 1985 by NEMA.

Table 6.4 Comparison of Specific Applications of Enclosures for Indoor Hazardous Locations

Provides a degree of protection against atmospheres typically containing (for complete listing, see article 500 of the National Electrical Code):	Class	Type of enclosure							
		7 and 8, class I groups[a]				9, class II groups[a]			10
		A	B	C	D	E	F	G	
Acetylene	I	X							
Hydrogen, manufactured gas	I	X	X						
Diethel ether, ethylene, cyclopropane	I	X	X	X					
Gasoline, hexane, butane, naphtha, propane, acetone, toluene, isoprene	I	X	X	X	X				
Metal dust	II					X			
Carbon black, coal dust, coke dust	II					X	X		
Flour, starch, grain dust	II					X	X	X	
Fibers, flyings[b]	III					X	X	X	
Methane with or without coal dust	MSHA								X

[a]Due to the characteristics of the gas, vapor, or dust, a product suitable for one class or group may not be suitable for another class or group unless so marked on the product.

[b]A class III, division 1 location is a location in which easily ignitable fibers or materials producing combustible flyings are handled, manufactured, or used. See Section 500-6(a) of the *National Electrical Code*.

Source: Reproduced by permission of the National Electrical Manufacturers Association from NEMA Standards Publication No. 250, *Enclosures for Electrical Equipment (1000 Volts or Less)*; copyright 1985 by NEMA.

government or private standards agencies on the use or characteristics of high-pressure, high-temperature washdown equipment. However, there is a common thread in the general practice within the industry. Typical equipment used in this industry produces a 15° fan-shaped spray from the nozzle at approximately 195°F and 1500 psi. The pressures and temperatures produced by such cleaning equipment have overstressed the seals of
many installed NEMA 4-rated industrial controls, leaving behind
a water-penetrated enclosure, a particularly bad problem for optical and electrical devices.

High-Pressure, High-Temperature Washdown:
    The enclosure under test shall not be penetrated by the 15°
    fanned spray from a nozzle at a range of 6 in. when held for
    5 s on each sealing joint. The nozzle temperature shall be at
    least 195°F, pressure at least 1500 psi, flow rate at least 4
    gpm, and the water shall have 0.05% wetting agent by weight.

6.3.2 Temperature

Temperature and humidity refer to the surrounding ambient air
conditions to which a photoelectric is subjected. Specifications
for these parameters are almost without exception given for the
ambient conditions of the air. However, at times the predominant
heat source may be radiant: from the sun, a furnace, or heating
elements. Radiant heat is difficult to quantify in the field; however, the resultant surface temperature of an enclosure may easily
be measured. The interior electronics are subject only to the case
temperature since it acts as a barrier to the outside air. As a rule
of thumb, the case temperature should not rise more than 10°C
above the sensor's ambient air temperature rating due to radiant
heating. Electronic components and other sensor construction
materials are capable of being stored safely at higher temperatures
than specified for normal operation, as there is no internal heat
generation during storage and materials can be assumed free of
mechanical stress at extreme storage temperatures.

Operating and Storage Temperature:
    The operating temperature range for industrial photoelectric
    controls and sensors shall be -20 to +70°C, as specified by
    NEMA ICS 2-229. Specifications shall apply over the entire
    operating range unless specifically stated otherwise. Output
    rating may be derated at temperatures over 55°C. The stor-
    age temperature range shall be at least -40 to +85°C.

Component performance will often vary with temperature. For example, an LED emits about 50% of the light at 70°C as it does at -20°C. Power switching components have reduced current-carrying capacity at elevated temperatures. Many other components also change their performance over temperature. Performance stability over temperature should never be assumed unless specifically stated.

6.3.3 Humidity

Relative humidity is a measure of the water vapor content of the air relative to the air's capacity to hold water vapor at a given temperature. At 100% relative humidity, the air is fully saturated with water vapor and condensation begins. Relative humidity is of interest for two reasons: its effect on circuitry, and its effect on optics. High humidity can create leakage paths between circuit components by freeing surface contamination ions to migrate. Surface contamination can result from condensing air pollutants, a fingerprint, or trapped ions in the circuit board material itself. Although these leakage paths may produce seemingly insignificant currents, the performance of high-impedance components may be altered in the presence of leakage currents. Detection threshold and timing logic circuits are two of the most susceptible to performance alteration. Potted and conformally coated circuits provide the best protection from condensation. Optical components do not mind humidity until water droplets form on the lens. Water droplets redirect and scatter the light collimated by the lens, resulting in reduced or nonexistent optical performance. Although an enclosure may have excellent seals, water vapor often finds its way in through conduit entries. Potted or conformally coated circuits are recommended if condensation within enclosures has been a problem. NEMA 4- and NEMA 6-rated industrial photoelectric controls are usually able to tolerate noncondensing 0.0 to 95% relative humidity.

6.3.4 Shock

There are two basic sources of shock: dropping the photoelectric during transportation, and the banging or colliding of objects with the photoelectric or its mounting. Shock is an acceleration impulse that places tremendous forces on an object over a very short period. The magnitude of a shock acceleration is inversely proportional to the amount of time it takes the object to stop as impact is occurring. A very hard object dropped onto a

concrete floor will not bend or compress on impact and will
generate high shock forces over a short interval. When this
force is concentrated over a small surface area, the stresses
often exceed the material strength causing permanent physical
damage. An object in foam packing will take a few more milliseconds to stop the fall, as the packaging foam is compressed while
slowing the object. In addition to reducing the shock forces
by spreading the impact over a longer interval, the packaging
spreads the forces over a larger surface on the object, relieving
concentrated stresses at the corners. The purpose of packaging
materials is damage prevention in shipment, particularly rough
handling and accidental dropping. A well-designed shipping box
will prevent accidental damage to the product when dropped on
any face or corner of the box from normal carrying heights onto
a concrete floor.

Shock, Drop:

A shipping package shall protect its sensor from sustaining
any physical damage when dropped onto a concrete floor from
a height of 48 in., impacting each edge, corner, and surface
of the shipping container.

Shock, Impulse:

An industrial sensor shall withstand three repeated 11-ms-
duration impulse shocks of at least $30g$ on each mechanical
axis with no sign of damage.

Impulse shock can be produced by maintenance personnel ham-
mering on the conveyor frame near a mounted photoelectric, objects hitting a railing as they are loaded onto a conveyor, logs
dropped against a backstop, or a forklift truck accidentally
smacking the frame on which a photoelectric is mounted. This
type of shock rarely damages the photoelectric housing. Rather,
it weakens cantilever-mounted components, eventually breaking
them loose. Transformers, relays, and large filter capacitors
are a few of the more susceptible components, although other
parts may also weaken and break. Once parts break loose from
their mountings, electrical or mechanical malfunction is generally
immediate. Shock may also be responsible for intermittent behavior. Electromechanical relays may momentarily change their
state under high shock as the armature spring force or coil magnet force is overcome by the shock force. Relays are rarely
rated over $10g$ of shock. Fully potted sensor construction has
the highest shock tolerance since components are supported
by the material in which they are embedded and prevent the

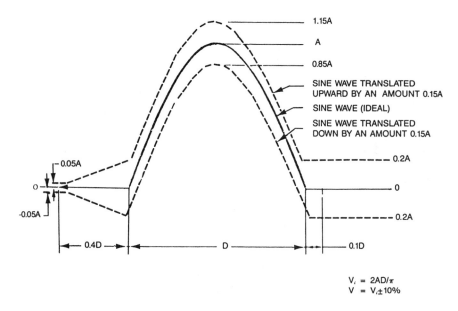

Figure 6.18 Half-sine shock pulse of acceleration amplitude A,
duration D, and velocity V. (Reproduced by permission of the
National Electrical Manufacturers Association from NEMA Standards Publication ICS 1-1983, *Standards for Industrial Control
and Systems*, copyright 1985 by NEMA.)

concentration of shock forces into fragile lead terminals. The
standard half-sine shock waveform shown in Fig. 6.18 comes from
MIL-STD-202F Method 213B and NEMA ICS 1-109. Test condition
J specifies a peak acceleration of 30g with a duration of 11 ms and
represents a minimum test for sensors and controls in typical industrial installations. Application-specific sensors designed for
light-duty environments may not require this level of shock tolerance. Heavy-duty environments may require as much as 100g.

6.3.5 Vibration

Many industrial machines and conveyors have motors, vibrators,
tumblers, or other equipment that produce constant vibration.

Vibration:

 An industrial sensor shall withstand repeated vibration sweeps
 from 10 Hz to 2 kHz at 10g acceleration or 0.060 in. displacement,

whichever is less, for a duration of 20 minutes on each mechanical axis with no sign of malfunction or damage.

Although vibration may be a problem at any frequency, its symptoms are particularly aggravated by vibrational frequencies that match component mechanical resonance frequencies. Vibration at the right frequency, even at low amplitudes, can excite a component to move violently in reaction to the vibration stimulus when the mass of the component body and spring constant of the mounting have the same natural resonant frequency as the stimulus. The resultant exaggerated movement at resonance produces great stress on the mounting and may eventually work-harden and break component leads or loosen mounting screws. These resonances are easily seen when stroboscope lighting is flashed at a slightly higher or lower frequency than the vibration frequency. Sources of mechanical vibration stimulus in an industrial environment rarely extends below 10 Hz or above 2 kHz. There are classes of photoelectric controls and sensors that are not intended to operate in high vibration environments. Examples include grocery checkout counters, toll gates, and hall security, where vibration is mild or nonexistent. However, even the lightest-duty sensors should be designed for at least $3g$ of vibration, while standard-duty sensors should be capable of withstanding as much as $10g$ of vibration over the frequency range 10 Hz to 2 kHz.

7

Selection Questions

You can't get there from here by guess and by golly.

Witatschimolsin

Now that you have reached this chapter, you are either a technical expert, confused by the facts, or starting here first hoping to take the shortcut by referring to the other material only as required. How you got here is not important as long as acceptable solutions are found. Selection is a process of elimination, judgment, and random discrimination; elimination of clearly unsuitable choices, relative value judgment to discard those of lesser performance, and random discrimination to reduce the number of equivalent solutions to a single decision. We cannot provide direct answers in this chapter, only a process, direction, and point of view to aid your decision process. There are a number of ways to approach a selection process. A simple decision tree that leads to an exact solution cannot be offered as a means to that end because of the endless variety in application detail and the tremendous variety of sensor available within the same basic classification. Often, however, the best way to free ourselves from analysis paralysis on a complex issue is to find out how similar problems have been solved by others, use that solution as a starting point, then modify the solution to fit the specific situation.

Chapter 8 application examples may be able to provide a starting point, then the questions and rules of thum of this chapter can be used as a checklist to flesh out the details of a final selection. There are five key questions that must be asked and answered to lead to a final selection decision, each of which is examined in this chapter.

1. What requires detection?
2. Where is it located?
3. How fast does it move?
4. With what will the photoelectric interface?
5. How much does installation and maintenance count?

7.1 WHAT REQUIRES DETECTION?

The first focus of attention must be the object or feature to be
detected. The two most important factors are its size and how it
interacts with light. The feature may be the entire object, part
of an object, a mark on an object, or a hole in an object. To
choose the best sensor we must understand how well the feature
reflects or transmits light from the sensing beam and what restrictions the feature size places on the sensing zone dimensions.

7.1.1 Size

Let's examine how the size of an object or feature affects the type
of optics required for its detection. Figure 7.1 shows the effect
on the signal received when an object crosses the effective beam
of a thru-beam or proximity sensor. Nearly 100% beam blockage
is required to reduce a thru-beam's received signal below threshold. However, an object intruding only a fraction of the way into a proximity sensor's effective beam may reflect sufficient signal for detection. A small white object 0.1 in. in diameter in a
0.3-in.-diameter beam requires an excess gain of greater than 10
for detection, while an excess gain of greater than 100 would be
required if the effective beam diameter were 1.0 in. Clearly, the
size of a proximity sensor's effective beam should not be much
larger than the object if the object is to be detected reliably.

   Generally, reflex and thru-beam sensors must have their effective beams essentially completely blocked to detect an object's
presence. However, small objects can be detected successfully
by reflex or thru-beam sensors with effective beams larger than
the object by using a low-contrast detection amplifier/demodula-
tor which adjusts the threshold near the operating signal level
so that slight changes rather than complete blockage of the signal produces an output. Apertures often satisfy the need for a
reduced-diameter effective beam on a larger sensors. Large objects also benefit from a small effective beam when improved position repeatability in switching is required. A small effective

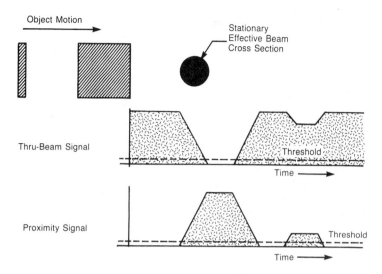

Figure 7.1 Effect of object size versus effective beam diameter on signal strength and threshold location.

beam implies a choice between focused proximity optics, miniature thru-beam sensors, thru-beam sensors with apertures, fiber optics, or a reflex sensor with a very small reflector. Small objects and effective beams usually imply the need for a faster response since an object is likely to spend much less time crossing through the effective beam.

Rule of Thumb:
1. The effective beam of a sensor should not be larger than the object or feature to be detected.
   Exceptions: a.   A proximity sensor with extremely high excess gain at the operating range is used.
   b.   Low-contrast detection circuitry is used.

7.1.2 Reflectivity/Transmissivity

Not only do objects have greatly differing reflectivities and transmissivities, but sensors also have quite different excess gain characteristics. Let's take a look at some of the reflectivity and transmissivity issues generally encountered.

In proximity detection, the excess gain of the sensor must be
great enough at the working range to compensate for the light
absorption of a low-reflectivity object, yet not so high as to detect background or peripheral objects. The vast majority of diffuse reflecting objects can be detected with proximity excess gains
of less than 20. The blackest diffuse reflecting objects rarely require excess gains of greater than 50. Polished metal and glass
may require excess gains over 500 to achieve detection with standard proximity photoelectrics unless the specular reflection from
the surface can be maintained to within about 3° of the beam axis.
Glass and polished metal are more easily detected with wide-angle
proximity sensors designed for this purpose which allow over ±30°
of alignment tolerance. In applications requiring the detection of
a color or shade contrast, too much excess gain will prevent detection. Here the excess gain must be adjusted high enough to
reliably see the lighter shade, but not the darker. A change
from infrared to a visible red or green LED source may be required to improve color contrast. Standard photoelectrics may
have trouble reliably detecting low-contrast marks, due to built-
in circuit hysteresis. Tougher applications may require a low-
contrast detection circuit or a color mark sensor.

Thru-beam sensors have tremendous excess gain at close range.
Thru-beam detection is particularly recommended in the most contaminated environments, where dirt or dust rapidly builds up on
the lens; where very long ranges are called for; or where internal contents of or markings on translucent objects are to be sensed.
Because of their high excess gain, thru-beam sensors can actually
see through many translucent objects that would normally be expected to block the beam. This capability is sometimes a solution
and sometimes is itself the problem. Selection of thru-beam sensors with low excess gain at the desired operating range may be
required to ensure that the object's signal transmission attenuation is not overwhelmed by surplus sensor performance. At close
ranges, smaller lensed thru-beam sensors are more easily aligned
and have lower excess gains, due to their inherent wider radiation
pattern and field of view than do their long-range larger-lensed
brothers. Detection of features with little difference in transmissivity, such as a seam on a plastic web, may require low contrast
or differential detection circuitry. Thru-beam sensors with analog outputs can be used for optical density measurement of gases
and liquids or location measurement of optical barriers.

Reflex sensors will sense through most transparent materials
but cannot sense through translucent materials because their excess gains cannot overcome the extreme attenuation of the signal

in a double pass through a diffusing medium. Polarization of reflex sensors eliminates false triggering caused by reflections from
specular reflecting mirror-surfaced objects. Mirrored surfaces
behind stressed plastics such as stretch wrap or injection-molded
plastics will negate the effectiveness of polarized reflex sensors.
Polarized reflex sensors significantly reduce the excess gain available from a standard infrared reflex sensor and are not recommended for applications likely to have any significant lens contamination.

Rules of Thumb:
    2. Proximity excess gain of 20 will detect most materials.
        Exceptions: a. Glass and mirrored objects
                    b. Velvet and carbon black objects
    3. Color-contrast discrimination can be improved by choice of
        source wavelength or use of low-contrast detection circuits.
        Exception: a. Carbon black marks are always great.
    4. Thru-beam sensors are the best performers in long-range
        and high-contamination environments.
        Exception: a. High-performance thru-beam sensors can
                        easily have too much punch at close range.
    5. Larger-lensed thru-beam sensors increase range and excess
        gain performance at the expense of alignment ease.
    6. Polarized reflex sensors eliminate false triggering from spec-
        ular surfaces at the expense of gain/range performance.
        Exception: a. Stretch wrap and injection-molded plastic
                        negate this advantage.

7.2 WHERE IS IT LOCATED?

Optimum photoelectric selection must consider the location of a
photoelectric relative to the object being sensed, adverse local
environmental conditions, and the location of surrounding objects.
Evaluation of these parameters is critical since sensing characteristics change radically with distance, sensor inability to tolerate
local environmental conditions will affect reliability, and objects
adjacent to the effective beam may alter the sensor's performance.

7.2.1 Distance

The purpose of the excess gain, field of view, effective beam, and
sensing zone specifications is to provide the sensor performance
information required for making selection decisions. Each of these
parameters must be individually evaluated in the selection process.

Although performance between various sensor designs varies
widely, a few generalizations can be made. For distances greater
than about 30 ft, the only choice is thru-beam. Below 30 ft, reflex may be considered in addition to thru-beam. Below about 6
ft, proximity may be considered in addition to reflex and thru-
beam. Because excess gain is a function of sensing distance,
operating at the limits of the specified range for a particular sensor will result in poor system reliability, particularly in a dirty
environment. Sensors should be selected so that the remaining
operating margin is at least a factor of 2 after accounting for
filthy conditions and dark targets. Conversely, thru-beam sensors rated for great distances may have excess gains so high that
at close ranges sufficient light will go through or be reflected
around the object by nearby surfaces rendering the sensor inoperative. For example, a thru-beam sensor rated for 400 ft will
have an excess gain of 160,000 at a range of 1 ft; excess excess
gain, indeed! .Except in special circumstances, a selected thru-
beam sensor's rated range should not exceed 10 times the intended
operating range.

The field of view, effective beam, and sensing zone are all
closely related; however, they do not vary identically with distance. Each describes the sensing field from a slightly different
point of view. The field of view simply describes where the
source light goes or where the detector is looking. The effective
beam describes the sensitive area of the beam: where the field of
view of the source and detector overlap. The sensing zone is a
description of the interaction between the effective beam, excess
gain, and a specified target. Unfortunately, it is.a rare occurrence to see more than one of these parameters specified in a
catalog or data sheet. However, a single specification combined
with the excess gain graph and understanding of the underlying
principles described in earlier chapters provides a basis for estimating the others. Many special sensors have been developed
to optimize the effective beam shape for particular application
classes. The effective beam diameter may take the shape of a
wide or narrow cone, a diamond-shaped dual cone, a section of
a pie, or a cylinder, as shown in Fig. 7.2. Generally speaking,
the cylinder shape belongs to both the reflex and thru-beam sensor styles, while the other shapes belong to proximity sensors.
These sensor characteristics must be matched to the understood
requirements of sensing distance, optical characteristics of the
object, and conditions surrounding the sensor which might otherwise interfere with detection.

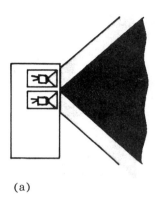

(a)

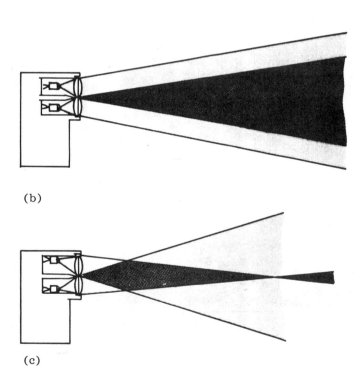

(b)

(c)

Figure 7.2 Effective beam shapes: (a) wide-angle proximity; (b) long-range proximity; (c) short-range proximity; (d) focused proximity; (e) thru-beam; (f) reflex; (g) curtain of light reflex.

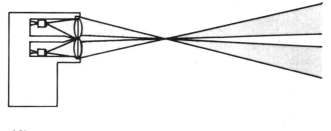

(d)

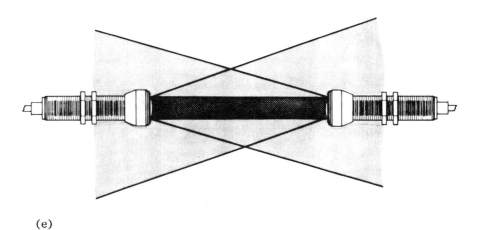

(e)

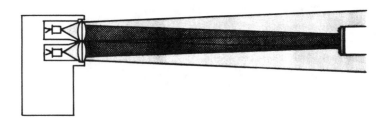

(f)

Figure 7.2 (continued)

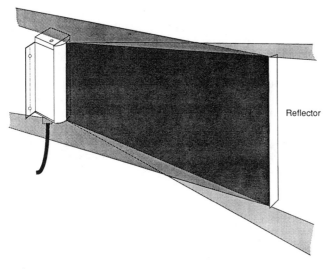

Reflector

(g)

Rules of Thumb:
   7. More excess gain at the operating range is better.
      Exceptions: a. Thru-beam sensors that see around or
                          through
                     b. Proximity sensors that see too far
                     c. Reflex sensors fooled by glass or metal
                     d. Low-contrast feature detection
   8. It is better to estimate the beam shape characteristics from
      sparse data than to pick a sensor blindly.

7.2.2 Local Environment

Most ambient lighting is not a problem for today's modulated photoelectrics; however, operation in the presence of direct sunlight requires at least 9000 fc of ambient light immunity outdoors and 5000
fc indoors where sunlight may enter through a vertically mounted
window. Unmodulated photoelectrics are restricted to operation
in fairly dark or shrouded environments or where its light source
is the only significant source of light.
   Most photoelectrics handle the 0 to 55°C (32 to 131°F) temperature range found in normal working conditions. Outdoor operation may easily hit -30 to -40°C in deep winter and 70°C or more

when baking in the summer sun. Radiant heat from the sun, a furnace, or an oven can raise the case temperature well above the ambient air temperature. The case temperature is always more important than the ambient air temperature. For operation above 70°C, glass/metal fiber optics and lens attachments can be special ordered to operate at temperatures exceeding 600°C.

The enclosure rating of a photoelectric must be checked for environmental compatibility. NEMA 1 enclosures should almost always be mounted in another industrial enclosure, NEMA 4 enclosures tolerate hosedown cleaning, NEMA 6 enclosures tolerate temporary submersion, and NEMA 13 enclosures tolerate oil or coolant splashes and sprays. Sealing out contamination is required for electrical connection safety, reliability, and is often required to prevent internal lens fogging.

Attack by solvents, acids, alkalis, and other chemicals is a problem particular to specific industries. Plastics, metals, and glass each have their individual chemical assailants. Generally, at least three different material types are exposed on a photoelectric: the lens, the body, and the mounting hardware. The photoelectric's construction material specification must be checked for compatibility with the known chemical environment of the installation. Appendix C lists commonly used plastic materials for photoelectric body and lens components and their chemical tolerances. Some manufacturers may be willing to accommodate changes in construction materials for a large-enough special order.

The electrical and radio noise environment must be considered from the start. Photoelectrics have varying degrees of susceptibility to both of these interference sources. Photoelectrics not specifying some immunity to these interference sources should be sample tested in the presence of a walkie-talkie and trashy power to verify compatibility prior to major commitments.

Rule of Thumb:
   9. The environmental condition not checked for compatibility is the one that will cause application failure with tangible and intangible consequences 10 times as costly as the current investment.

7.2.3 Mounting Location

There are three basic mounting location configurations: panel mounted with remote optics, frame mounted with remote optics, and direct mounted at the sensing location. Figure 7.3 illustrates each of these mounting styles. Panel mounting is chosen to centralize controls that must interact with each other as a system, or

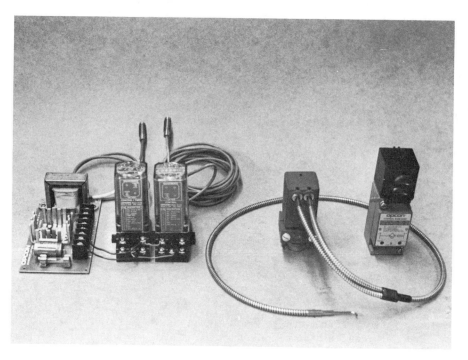

Figure 7.3 Mounting options include (from left to right) panel mounted with remote optics, frame mounted with remote optics, and direct mounted with self-contained optics.

to restrict access by line personnel to control adjustments. Panel mounting provides systems integraters with the freedom to assemble and test a control system prior to installation, and provides protection of the ensemble of system control components, some of which may not have enclosures rated for industrial or outdoor environments. Photoelectrics can be direct mounted at the sensing site when sufficient space is available and the enclosure is rated for the environment. However, mounting space is almost nonexistent in some machinery. When self-contained 18-mm tubulars and other miniatures are too large, the alternative is a frame-mounted photoelectric with remote optical heads or fiber optics. This style of photoelectric has a separate control unit that is mounted on the frame of the machine or conveyor while the optical sensor is remote connected to the sensing site. Remote optical head and fiber optic photoelectrics cost a little more but

offer quite an array of control and optics options. The price generally paid for size reduction is optical performance. Fortunately,
where space is a premium, required sensing distances are usually
short.

Mounting location also concerns the ruggedness of the packaging. Certain mounting locations may make a photoelectric quite
susceptible to physical abuse. It is not uncommon for a 200-lb
electrician to use anything available, including a photoelectric,
as a step stool when climbing around to do his job. I've even
done so. Protruding photoelectrics are likely to be bumped by
carts, boxes, elbows, and other objects. A photoelectric package intended for frame mounting or direct mounting at the sensing
site ought to be stout enough in construction and mounting to take
such abuse.

Rule of Thumb:
   10. A photoelectric package that can be bumped or used as a
       step, will be.

7.3 HOW FAST DOES IT MOVE?

What is the minimum time available to the sensor to respond to an
object's presence in the sensing zone? What is the maximum number of objects per second that can be sensed? These are the bottom-line speed questions. Response time is simply the length of
time it takes for a sensor's output to respond to an instantaneous
change from a dark-to-light or light-to-dark condition at the detector. Response time is directly related to the maximum possible
number of sensed objects per second. To sense an object's passing, the sensor must first respond to its presence and then to its
absence. Only then can it be prepared to start detection of the
next object. Let's say that a particular sensor responds to an
object's presence in 4 ms, and its absence in 6 ms. Then the
minimum cycle response is the sum of the two, or 10 ms. Since
10 ms is 1/100 s, the maximum possible number of objects that
could be sensed per second would be 100. Response time is,
however, complicated by related factors, such as effective beam
width, signal strength, output device response time, and repeatability. Let's examine these one at a time to clarify their rele-
vance.

7.3.1 Object Transition Time

Transition of an object through the effective beam width is important for both response time and repeatability. The leading

and trailing edges of an object moving at 25 in./s through an effective beam 1/4 in. in diameter will cross the beam in 10 ms. Where in the beam crossing does detection occur? If we know how much excess gain the system has, we can calculate what percent of the beam must be crossed for detection. But how much excess gain will be left after dirt builds up, and how far will the detection point move? Clearly, the larger the excess gain, the smaller the variation in switch point with dirt buildup, aging, and temperature effects. Figure 7.4 shows how the response time of an instantaneous ideal photoelectric sensor will vary as lens cleanliness degrades. If better than 10 ms response time or long-term repeatability is to be achieved, there are two choices: Be sure the excess gain is always very high, or choose a sensor with a smaller effective beam width. A slot or hole aperture over

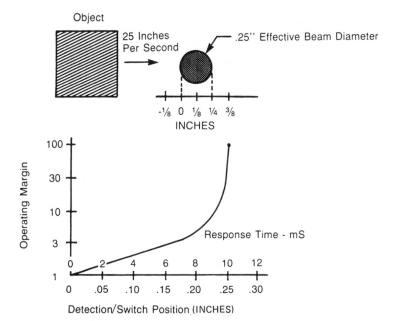

Figure 7.4 Response-time variation and detection point shift as a function of change in operating margin due to dirt buildup on the lens. Low optical operating margin results in high sensitivity to dirt.

the lens reduces the effective beam width for thru-beam sensors.
A narrow piece of retroreflective tape or a masked retroreflective
disk can control the effective beam width for a reflex sensor.

Rules of Thumb:
11. High excess gain and a narrow effective beam is required
    for minimum response time and maximum repeatability.
12. Applications requiring very small effective beams gener-
    ally also require sensors with faster response times.

7.3.2 Sensor Response Time

Response time may not be the same for both light-to-dark and
dark-to-light transitions and may vary with the level of excess
gain. When the excess gain is very low, approaching unity, the
dark-to-light response time increases and the light-to-dark response time decreases. This is an artifact of analog integrating
detectors used in most photoelectric designs. Selection of a
photoelectric that meets the optical requirement marginally is
likely to give less than marginal satisfaction. Photoelectric sensors should not be operated near unity excess gain.

Fast response time requires a solid-state DC interface. NPN,
PNP, and FET transistor output circuits are very fast reacting,
usually conveying the detection circuit's decision to the output terminals in less than one hundredth of a millisecond. Transistor outputs do not add measurable response delay to a sensor.
Optically coupled FET outputs circuits usually add 1 to 5 extra
milliseconds of response delay as a result of the optocoupler. Relays and AC circuits add many milliseconds of switching delay to
a sensor's output. Relays typically take 5 to 15 ms to open or
close contacts after the detector changes state. AC circuits are
stuck with the cyclic nature of the AC power. There is about a
3-ms period once every 8.33 ms when the 60-Hz AC voltage goes
through gentle zero crossing. An output switched on during
this 3-ms interval cannot provide sufficient energy to the load
to start a perceptible response. The switching technology used
on AC output circuits also has a profound effect on the response
time. Bridge-rectified NPN transistor designs switch immediately.
Optically coupled FET designs delay switching a few milliseconds.
Triac and SCR output designs do not turn off directly, but wait
until the next zero current crossing, which may be from 0 to 8
ms away. Some triac and SCR designs will turn on immediately,
while "zero crossing" designs wait for the next zero voltage crossing following the trigger before turning on. The minimum AC
switching response time is at least 3 to 8.33 ms and cannot be

repeatable under 3 ms and often not under 8.33 ms. Even though the sensor demodulator may react very quickly, AC-powered outputs cannot be relied on for quick responses.

Rule of Thumb:
13. Response times of less than 10 ms require use of a DC solid-state output interface.

7.3.3 Repeatability

Repeatability is often more important than response time. A typical photoelectric source LED produces 1000 or more pulses per second. Although its specified response time may be many milliseconds, its basic repeatability is likely to be ±1 ms or better. For example, it may take five pulses or more to change the detector's state; however, the repeatability of the change will probably be ±1 pulse. The point is that there is a distinction between timing repeatability in a process and object sensing rate. Two-millisecond repeatability is easily achievable from a garden-variety 10-ms-response-time photoelectric. Response time should not be confused with its cousin "repeatability," both of which are significantly affected by excess gain, the physical travel speeds across the effective beam, and the delay inherent in AC output circuits.

Rule of Thumb:
14. Sensor repeatability is always better than worst-case response time.

7.4 WITH WHAT WILL THE PHOTOELECTRIC INTERFACE?

There are two interfaces of concern: the load and the power supply. The selection decision is often quite clear after consideration of required response time, preexistent load requirements, available power, and cost.

7.4.1 Load Interface

A photoelectric sensor or control is not successful if it cannot cause some external effect as a result of its sensing task. For this reason, selection by load interface is prioritized above selection by power supply. Reporting of the sensing event to another device which ultimately takes the final action is the job of a "sensor" with a simple light-duty output interface. Local direct control of physical external devices which take the final action is the job of a "control" with medium- to heavy-duty output interface and optional logic capabilities. The boundary between sensors and controls can be quite fuzzy at times and is often further confused by industry's inappropriate use of the term "scanner" for sensor or control. Sensors interface to AC, DC, and analog programmable controller input circuits, optical isolator modules, logic-level circuits, counters, and analog servo control circuits. Controls interface to motors, brakes, solenoids, ejection devices, contactors, heaters, and lamps. The output circuit interface is chosen to match the requirements of the load circuit. The parameters to be investigated are response time, AC versus DC, off-state blocking voltage and leakage current, on-state current rating and voltage drop across the switch, and capability of AC interfaces to handle low-power-factor loads. Selection is determined primarily by output-response-time requirements, load interface characteristics, available power compatibility, and installed cost. Many photoelectric controls have plug-in output device options that provide choices between relays, solid-state AC, or solid-state DC interfaces.

Timing logic must be selected for a control if the sensing and output timing are not identical. Although on/off time delays, one-shots, retriggerable one-shots, delayed one-shots, on-delay one-shots, shift registers, divider/counters, under/overspeed detection timers, and other timing logic are readily available options, only a subset of these is generally available in any given product family. Consequently, the availability of the desired timing function may provide the greatest solution limitation and become the highest priority in the selection process. However, when the control system is panel mounted, the addition of timing logic external to the photoelectric provides maximum selection flexibility, making it an attractive alternative to photoelectric controls with built-in timing modules.

Rule of Thumb:
15. The load interface is most often a prescribed selection, dependent on other system requirements and choices.

7.4.2 Power Supply Interface

The power supply interface choice is most often subordinate to the output interface choice. Large control systems are generally designed to reduce the number of overhead system components, particularly extra transformers or DC power supplies. The best option for a sensor's power supply is to use the power

that is most readily available. Simple sensors almost always require that power and load interface supplies be the same (i.e.,
both AC or DC, but not one of each). Modular photoelectric controls with isolated plug-in output devices generally allow the mix
and match of AC and DC supplies and outputs. Although AC is
still the preferred interface today, the advantages in cost, response time, and safety of DC sensors, together with the growth
in solid-state control systems requiring DC interfaces, is increasing the market preference toward the DC supply and load interface.

Rule of Thumb:
   16. First find a sensor that will do the required detection,
        then match it to the output requirements, and finally,
        worry about power supply options.

7.5 HOW MUCH DO INSTALLATION AND MAINTENANCE COUNT?

Installation and maintenance are often ignored as serious considerations in the selection process because the system designer or
OEM is either not responsible for these items or not aware of the
cost implications. What is the added cost of installation when
each of 500 photoelectric controls requires an extra 5 minutes to
assemble and mount? What is cost of lost production when an
automotive assembly plant is down an extra 5 minutes for a difficult replacement of an inoperative photoelectric?

7.5.1 Mounting

Mounting costs include the cost of custom mounting brackets and
the labor costs to drill holes, install, and align the photoelectric.
The existence of a suitable mounting bracket can save a lot of
time and grief by eliminating the need to customize an unsuitable
design to meet the requirements. A good mounting bracket provides adjustment freedom to accommodate the optical alignment
needs and minimizes the number of holes to be drilled and screws
to be fastened. There can be a significant difference in mounting difficulty or annoyance between two similar photoelectrics.
Although it is not difficult to drill a few holes for mounting screws,
the fewer the number of screws, the quicker the job is done, particularly when any quantity is involved. The difference in mounting difficulty between two similar photoelectrics may be as much

as two screws versus six screws using the manufacturer's supplied bracket!

A few standards for mounting have emerged over the years.
These standards have been pushed primarily by the automotive
industry for its own needs. The two primary standards are the
mechanical limit switch style shown in Fig. 7.5a, and the 18-mm
tubular style shown in Fig. 7.5b. Other packaging styles have
also come to be at least nearly standard in shape and in mounting. These include the "flat pack" in Fig. 7.6a, the "fire hydrant" in Fig. 7.6b, and the "little guy" in Fig. 7.6c. These
standards most often are the result of manufacturers following
the lead of a product that has captured market share in the hopes
of participating in the maintenance and repair market or providing
a lower-cost, better featured, or more reliable alternative to the
market leader. Replacement of a photoelectric with another brand
may not be a trivial task if it does not fit directly in place of the
malfunctioning unit, both mechanically and electrically.

Installation in certain mounting locations requires the addition
of a few precautions. Radiant heat from furnaces, ovens, hot
steel, and other radiant sources can raise the case temperature
of a photoelectric considerably above its rated operating temperature range, in spite of the lower ambient air temperature. Use
of a heat shield in front of the photoelectric, as shown in Fig.
7.7, with a hole for the beam to peer through will significantly
reduce the radiant heating of the case. The shield will also reduce the radiant heat exposure to the lens due to blockage of
radiant energy arriving off-axis.

Installation in hazardous environments requires the use of explosion-proof enclosures or intrinsically safe sensors. Operation
of any electrical equipment in the presence of explosive gases,
grain dust, volatile paint or solvent sprays, and other flammable
materials should be approached with great caution using industry
standard methods of safety that comply with applicable codes.
Some photoelectric manufacturers make available explosion-proof
enclosures and intrinsically safe barriers with approved sensors,
or will recommend sources for them. One of the natural virtues
of fiber optics is its ability to have the electrical control in a safe
remote location while the optical fibers do their duty in the hazardous sensing location.

A common problem arises when a photoelectric is mounted be-
neath and looking up between conveyor rollers and is expected
to sense objects on the far side of the rollers without sensing
the rollers themselves. The worst way to mount the photoelectric is shown in Fig. 7.8. The normal effective beam slips nicely

(a)

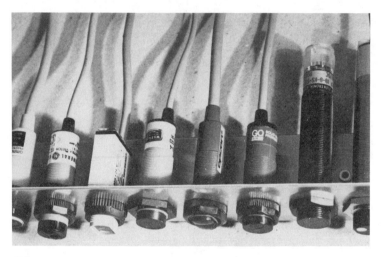

(b)

Figure 7.5 Industry standard packaging styles: (a) mechanical limit switch; (b) 18-mm tubular.

between the rollers as the LED radiation pattern and detector field of view overlap. However, the rollers create a secondary effective beam from the double reflection and lateral shift of the

(a)

Figure 7.6 Near-industry standard styles: (a) flat pack; (b) fire hydrant; (c) little guy.

beam that results in a sensor that sees the rollers just fine all the time. There are two methods to remedy this situation. First, tilt the sensor as shown in Fig. 7.9 so that the specular reflection from the rollers is not returned to the photoelectric. This method, however, can be defeated by paper or other material stuck on the rollers which may provide sufficient returned signal via a double diffuse reflection. Second, use a sensor that has optics no larger than 80% of the roller gap dimension and a field of view narrow enough to skip between the rollers without touching.

System reliability will be highest when the mechanical, optical, and electrical integrity of the design each best match the application requirements. Vibration, electrical noise, chemical attack, washdown, dust and dirt, and operating temperature all may have significant effects on system reliability. Never assume that a sensor or control can handle any environment; check it out in the product specifications.

(b)

(c)

Rules of Thumb:
> 17. Mounting costs are increased at least one dollar per ex-
>   tra screw required.
> 18. Hours of trouble can be saved by spending minutes to
>   understand special mounting requirements prior to the
>   installation.

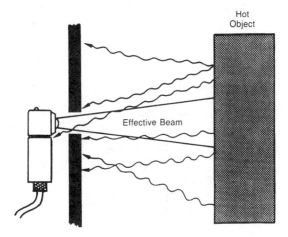

Figure 7.7 Heat shields block radiant heat and keeps the enclosure temperature within operating specifications.

7.5.2 Wiring

The cost of wiring is one of the key installation considerations. Photoelectric installation costs include labor and additional ancillary hardware required to make it work. If a cable is provided, is it long enough, or must a junction box be installed to make a splice to lengthen the cable? Can the cable be ordered to the required length from the manufacturer? An integral cable may save cost until it is not long enough; then splices and junction boxes must be added. How many wires must be pulled through conduit to a photoelectric with terminal blocks? Some industries estimate the cost to pull wire is at least one dollar per foot per wire, and this does not include the cost to install conduit. The photoelectric with the lowest wiring cost is the two-wire style. The photoelectric style with the highest wiring cost is the long-range thrubeam, since wiring is required in two locations.

When a photoelectric does fail, how quickly can it be replaced in order to reduce production downtime? Replacement difficulty is proportional to the number of mounting screws to be removed and whether or not wires must be disconnected or pulled. Many photoelectrics have been designed to permit simple plug replacement of the electro-optical components onto a wiring base that remains permanently mounted. Photoelectrics with a cable connector

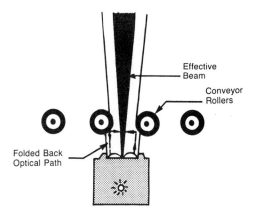

Effective Beam

Conveyor Rollers

Folded Back Optical Path

Figure 7.8 A poorly mounted photoelectric may sense the rollers rather than objects beyond.

or a connectorized mounting base with terminal blocks eliminate the need to rewire the photoelectric during replacement. Rewiring results in considerable lost time during the change, and potential rewiring errors.

Industrial wire trays and conduit are a significant source of electrical noise. Fast edge waveforms from high-frequency switch mode drives and the arcing of contacts making or breaking inductive circuits are easily coupled into wires and cables lying in close proximity. Although pulse-modulated photoelectric sensors and controls are designed to reject extraneous signals, there are limits to what can be expected. Running the power supply wires in such trays and conduits should not be a problem for well-designed photoelectrics. However, if detector cables of remote sensor photoelectrics are run in the same conduit with high-voltage trashy power lines, noise trouble can be expected. Typical power device control lines switch currents on the order of 10 billion times greater than the detector currents produced by a remote photodetector. That is why remote photoelement cables should always be installed in separate trays or conduit.

Rules of Thumb:
   19. Connectorized or plug-replaceable photoelectrics result in lower long-term system cost when downtime and maintenance are considered, particularly in physically abusive environments.

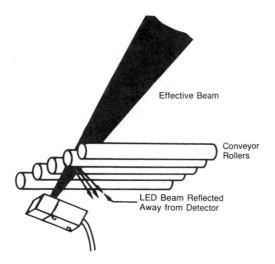

Figure 7.9 A tilted photoelectric eliminates the specular return from the rollers.

20. Photoelectric remote sensor head cables should not be put in the same conduit with power switching circuits.

7.5.3 Alignment and Adjustment

Once the photoelectric is mounted and powered up, it may be aligned optically with the target. For proximity sensing, the object to be detected is moved through the sensing zone repeatedly while the sensor is mechanically adjusted to provide the correct lateral sensing location, and excess gain is electrically adjusted to result in the highest operating margin or optimum threshold level in low-contrast applications. Reflex sensors are aligned by mounting the retroreflector in the intended location and loosely mounting the sensor. The reflex sensor is then tilted up and down and right and left as shown in Fig. 7.10 to determine the direction that best centers the beam on the retroreflector. Alternatively, the reflex sensor may be securely mounted first, followed by a beam center search similarly executed by moving the retroreflector. It is a rare circumstance when the excess gain is not adjusted to the maximum on a reflex sensor. Exceptions include applications where the blocking object is semitransparent or where weak inadvertent specular reflections are to be

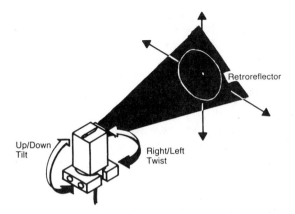

Figure 7.10 Optimum alignment is obtained by locating the effective beam center on the retroreflector through an up/down and right/left search with the sensor direction or reflector position.

ignored. For thru-beam sensing, the source and detector are loosely mounted so that the effective beam between them will create the sensing zone in the proper location. The individual heads are then adjusted mechanically to point in the apparent optimum direction until detection is achieved. Then each head is individually mechanically adjusted for the optimum pointing direction by moving the sensor's pointing direction up and down and right and left to determine the center position in each direction. The source and detector are then tightened down in these positions. Excess gain is adjusted high enough for reliable operation, but not so high as to result in transmission through the object or secondary transmission paths around it.

Some manufacturers provide built-in alignment aids to reduce the anxiety involved in sensor alignment. Reflex, focused proximity, and thru-beam sources with visible beams improve alignment ease since the beam location can be determined directly by observing the spot created on the target or retroreflector, or by directly looking back toward the sensor and finding the place where the light appears brightest. Some photoelectric receivers have built-in alignment indicators that provide analog signal strength feedback by one of several methods: the blinking rate of the indicator LED, changing the indicator LED brightness, or changing position on a built-in LED bar graph.

After the photoelectric control is installed, adjustments are made to delay or one-shot modules to produce the desired output

timing. Timing circuits should be operated at the upper end of
timing ranges so that better adjustment resolution is possible.
Adjustment of timing circuits operated at the lower end of the timing range can be quite tedious, if not nearly impossible. Timing
circuits and line speeds are susceptible to variation with time,
temperature, supply voltage, and unauthorized tampering by un-
solicited helpers. The best way to stabilize delay timing is to use
a pulse generator tied to the conveyor as a timing base for a shift
register. The result is an output delay set by the number of
pulses counted and is not susceptible to time, temperature, supply voltage, or line-speed effects.

Rules of Thumb:
    21. Alignment aids reduce installation time and anxiety and
         reduce the likelihood of future maintenance problems.
    22. Shift register delays eliminate output timing errors due
         to timer temperature dependence and line-speed varia-
         tions.

7.5.4 Upkeep

The cost of upkeep is generally limited to keeping the lenses,
reflectors, and fiber optic tips free from dirt and water accumulation. The interval between cleaning will depend on the cleanliness of the environment and the optical operating margin available. High excess gain generally reduces the maintenance required in filthy environments and improves the operating margin
and reliability. One successful method of combating dirt and dust
buildup is to surround the lens with an aperture fitted for shop
air that maintains a steady flow of air outward from the lens so
that falling sawdust, coal dust, lint, sprays, and other contaminants are continually blown away from the photoelectric lens, as
shown in Fig. 7.11. Lens fogging can be prevented by coating
the lens with the same compound as that commonly applied to eye-
glasses for that purpose. Keep in mind next time, when on a
cold winter morning, the plant manager asks you to go to the
dock to check your reflexes, he probably isn't worried about
your health. He's referring to the fogged up reflexes on the
shipping dock door control. Preventive maintenance is the key
to system reliability in environments with high contamination. The
alternative to automatic maintenance of optics is regular attention
by the maintenance crew. Maintenance costs are paid at installation, during added maintenance personnel tasks, and as factory
downtime.

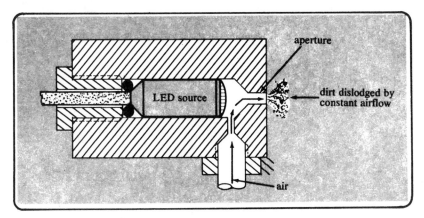

Figure 7.11 Air-purged apertures eliminate regular lens cleaning maintenance in contaminated environments. (Courtesy of Opcon, Inc.)

Rule of Thumb:

23. Installation and maintenance must be among the principal considerations in a successful photoelectric selection process. Total cost is both a short-term and a long-term measurement.

8
Application Examples

Fortune favors the brave

Terence
Phormio, Act i, sc. 4

I have selected these application examples for their instructive
value, ideas portrayed, and techniques used in solving many typical sensing and control problems. Many of the clever techniques
and applications shown have value in related applications, with
slight modifications. The application notes have been organized
into five groups, to simplify digestion. Through the courtesy of
many photoelectric manufacturers, these previously published application notes are reproduced here. A manufacturer's specific
model number is indicated in many of them. In some cases it may
not be possible to implement the application with other equipment.
In other cases different equipment may suffice since the application is rather generic and numerous makes and models of similar
photoelectric equipment will perform well.

The bottom-line objective is to be successful. If the application information you need is not found in this book, the next re-
source available to you is the experience of numerous application
engineers. They are accessible through a phone call to a photoelectric manufacturer. Success does not require knowing all
things. It only requires knowing where to get information and
a bit of common sense in using the information acquired.

8.1 MATERIAL AND PROCESS CONTROL

8.1.1 Broken Yarn Detection

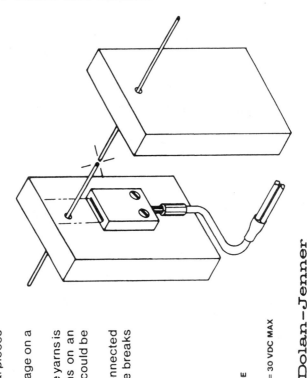

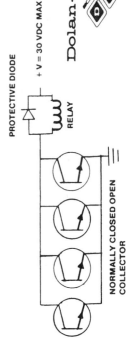

APPLICATION: To detect broken yarn

OBJECTIVE: To sound an alarm if one of several pieces of yarn is broken

DESCRIPTION: The control senses yarn breakage on a yarn spinning machine.

Four yarns are being spun into one. Each of these yarns is monitored by a Led-Pak IV control, which turns on an alarm if any breakage is detected. The yarn size could be as small as 0.005" (0.1 mm) in diameter.

The outputs of all Led-Pak IV controls could be connected in parallel for an "or logic" so that if one or more breaks occur, the alarm will be turned on.

EQUIPMENT:

4 - 2000 Base
4 - 2010 Electronic Module
4 - 2400 Input/Output Module
4 - EF824 Fiber Optics

8.1.2 Broken Thread Detection

BROKEN THREAD DETECTION ————————

MODEL #	DESCRIPTION
8171B	High-Sensitivity Control Unit
8215A	Low-Contrast Logic Module
1173A-100 (2)	Moderate-Range Thru-Beam Source
1273A-100 (2)	Moderate-Range Thru-Beam Detector
8530A	DPDT Relay
6143A (4)	Mounting Brackets

A pair of remote thru-beam sensors scan over and under multiple strands of thread. If a thread breaks and passes through one of the beams, the low-contrast logic module detects the sudden change in signal strength and energizes the output. Because this logic module does not react to slow changes in signal strength, it can operate in a dusty environment with little maintenance.

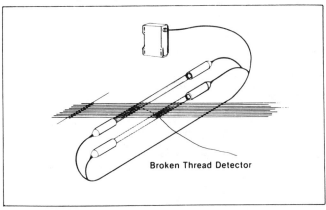

Broken Thread Detector

opcon®
sensing your needs

8.1.3 Broken Wire Detection

MAGNET
WIRE BREAKAGE CONTROL

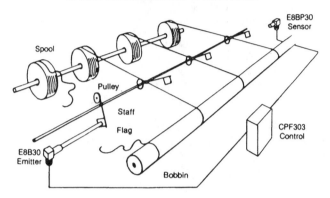

Magnet wire is subject to breaking as it is being wound on spools. Breakage must be detected and located among many spools instantly so that it can be repaired and the winding operation resumed. The problem is that magnetic wire itself does not offer enough obstruction to block or restore the light beam of a Photoelectric Control. The problem was solved when the above application was designed to control magnetic wire breakage. The magnet wire from spool to each bobbin passes over a grooved wheel which is mounted on a pivoted staff with a flag at the other end. The wire pulls the wheel down and elevates the flag. When a magnetic wire breaks, the flag drops and blocks the light beam.

8.1.4 Web Loop Control

WEB LOOP CONTROL ──────────────────

MODEL #	DESCRIPTION
8172A	Analog Control Unit
1471A	Reflex Curtain-of-Light Sensor
8272A	Analog Isolation Module
6210A	Retroreflector

A sensor that generates a "curtain of light" detects the length of a loop on a web drive system by measuring the amount of light returned from an array of retroreflectors. With this information, the analog control unit instructs a motor controller to speed up or slow down the web drive.

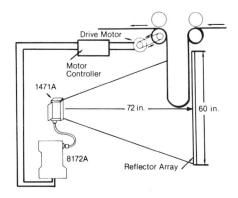

8.1.5 "Rate of Feed" Tension Control

"RATE OF FEED"
TENSION CONTROL

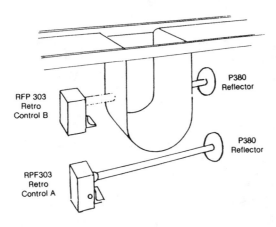

     Rate of feed is regulated by position of "loop" in relation to Photoelectric Controls A and B. When loop gets too short, the light beam of B is restored and Photoelectric Control B actuates, speeding up the feeding mechanism until light beam is again interrupted. When loop gets too long, it interrupts light beam of A and Control A actuates, slowing down the feeding mechanism.

8.1.6 Web Material Centering

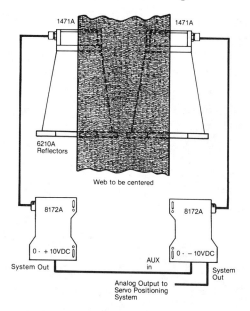

The lateral position of a traveling web is centered
by the differential balance between two analog
output photoelectric controls. The analog output
will cause the servo positioning system to maintain
both reflectors equally blocked by the web.

8.1.7 Web Edge Control

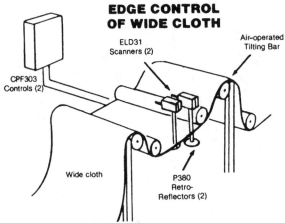

EDGE CONTROL OF WIDE CLOTH

ELD31 Scanners (2)

Air-operated Tilting Bar

CPF303 Controls (2)

Wide cloth

P380 Retro-Reflectors (2)

The above sketch illustrates a Photoelectric edge control for wide cloth going from a table into a sponge machine. The sketch shows the cloth in the correct position with both photo controls inoperative. Photoelectric monitoring consists of two scanners and two retro-reflectors positioned above and below one edge of the cloth to permit instant detection and correction when the cloth moves out of alignment. If the cloth interrupts the light beam at the right, the control signals an air mechanism to tilt the adjusting rod up at the right end. When the cloth moves over, the light beam is again restored. If the cloth lets the left light beam strike the retro-reflector, the control bar is tilted up at the left end to correct alignment.

8.1.8 Web Break Detection: Clear Plastic

CLEAR PLASTIC WEB BREAK DETECTION ——————

MODEL #   DESCRIPTION
8880C-6501   High-Current Control Unit
1380B   Short-Range Proximity Sensor

The clear web is detected by an extremely sensitive proximity control. Its short detection range makes it immune to reflective objects in the background. The extremely high excess gain helps it ignore reflection caused by fluttering of the web.

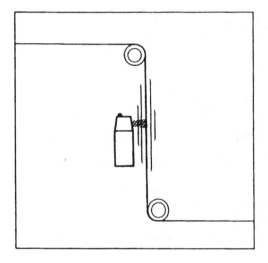

8.1.9 Web Break Detection: Paper

Paper Web Break Detection

A Reflective/Proximity Photoscanner senses a
web break and automatically signals the press
to shut down. This prevents the rollers from
becoming loaded with ink.

8.1.10 Web Splice Detection

WEB SPLICE DETECTION ─────────────────

MODEL #	DESCRIPTION
9072A	Differential Control Unit
8213B	One-Shot Logic Module
8526A	DPDT Relay
1173A-300 (2)	Moderate-Beam Angle Source
1273A-300 (2)	Moderate-Beam Angle Detector

When the two thru-beam detectors see the same signal strength, the output is zero. When the opacity of the web changes, as in a splice, the signal strengths are thrown out of balance and the output is energized. This system can be used on webs of different colors and opacities with no system reconfiguration.

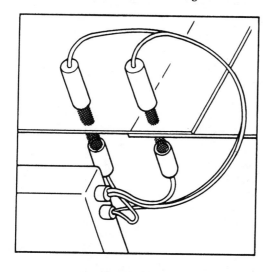

8.1.11 Web Flaw Detection

WEB FLAW DETECTION

Application: Find defects or holes in continuous web, provide output to stop web prior to next step in process.

Opcon products used: 70 Series.

Model #	Description
8171B	Control Unit
1471A	Curtain of Light
8215A	Low contrast module
8526A	Relay
6210A	Reflector

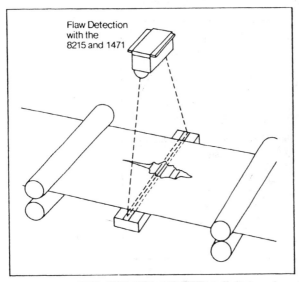

Flaw Detection with the 8215 and 1471

opcon®
sensing your needs

8.1.12 Roll Wind Length Control

AUTOMATIC WINDING
TO EXACT PAPER ROLL LENGTH

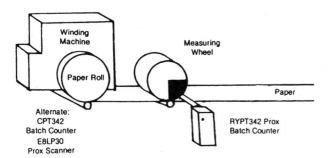

This Photoelectric Control application measures paper automatically as it is wound on a roll. The measuring wheel is 12″ in circumference and is painted ¼ black and ¾ white, as shown in the above illustration. The measuring wheel also runs free at speeds up to 1200 feet per minute. The contrasting black and white of the measuring wheel trips the high-speed photoelectric batch counter on which the required roll length is preset. After sufficient revolutions of the measuring wheel have indicated to the counter the completion of a pre-set roll length, the counter stops the winding machine.

8.1.13 Roll End Detection

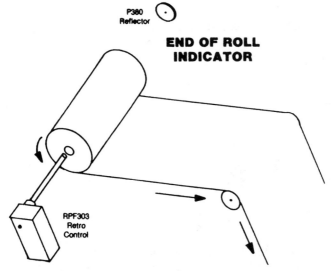

When an uneven rate of usage or frequent inter-
ruptions make it difficult to anticipate the need for a
new supply of material, Photoelectric Controls can alert
operators to shortages before they interfere with
automated production. In the above application, a control and reflector are mounted on opposite sides of a
paper roll with the light beam aimed slightly above the
core. When the paper is about to run out, the light beam
is restored and the control shuts down the machine.
The control can also sound an alarm, leaving a
predetermined amount of paper (or other material) on
the roll, so work in progress can be completed before
the depleted roll is replaced.

8.1.14 Trim Length Control

TRIM LENGTH CONTROL

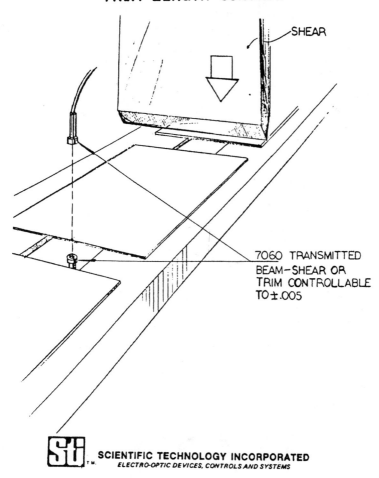

SCIENTIFIC TECHNOLOGY INCORPORATED
ELECTRO-OPTIC DEVICES, CONTROLS AND SYSTEMS

An STI 7060-series sensor is arranged a preset distance from a shear. When the beam is broken by the material to be sheared, the cutter is automatically actuated, cutting the material to the proper length. The sensor may be arranged on an adjustable yoke to permit changes in lengths. (Courtesy of Scientific Technology Incorporated.)

8.1.15 Filter Paper Length Control

FILTER PAPER LENGTH CONTROL ————————

MODEL #	DESCRIPTION
8171B	High-Sensitivity Control Unit
9082A	Fixed-Focus Proximity Sensor
8573A	Triac Output Relay

A fixed-focus proximity control with a triac output module interfaces with a programmable controller to measure a specific length of corrugated automotive filter paper. The control detects the presence or absence of a corrugation. When a predetermined number of corrugations has been detected, the programmable controller directs a shear to cut the paper.

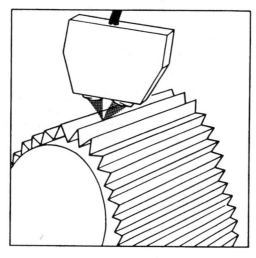

8.1.16 Cutoff Saw Control

CUT OFF SAW CONTROL ————————————

MODEL #	DESCRIPTION
1190A	Thru-Beam Source
8891A	Power Base
1290A	Thru-Beam Detector
8890A	Power Base
8250B	Light/Dark Logic Module

Note: All products listed are required for each two-foot increment.

An array of thru-beam controls detect the length of the log in standard two-foot increments. The correct saw is then activated to cut the log at its longest standard length. High optical performance is a must in this dusty and dirty environment.

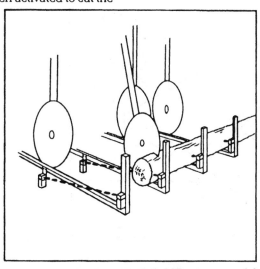

8.1.17 Hypodermic Needle Quality Control

HYPODERMIC NEEDLE QUALITY CONTROL ─────────

MODEL #	DESCRIPTION
8171B	High-Sensitivity Control Unit
1173A-300	Moderate-Beam Angle Source
1273A-300	Moderate-Beam Angle Detector
8573A	Triac Output Relay

A remote source and detector pair inspects for passage of light through a hypodermic needle. Their small design and stainless steel, waterproof housing are appropriate for crowded machinery spaces and frequent washdowns. High signal strength allows quality inspection with hole sizes down to 0.007 inch.

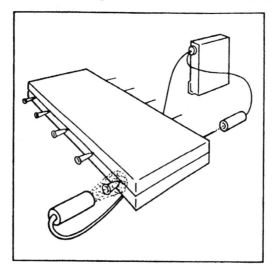

STEEL STRIP MISALIGNMENT DETECTION

A pair of bifurcated fiber optic cables is used with a single sensor to detect misalignment of a steel strip and to activate an emergency stopping mechanism. Note how the source signal splits into two bundles of fibers. A second bifurcated cable detects light from both sources and routes it back to the control unit. If either effective beam is completed, the output turns on and actuates the stopping mechanism.

Parts Required

Qty.	Model Number	Description
1	1550A-6511	110VAC sensor with cable
2	6222A-6502	Bifurcated fiber optic cables with smooth tips
1	8532A-6501	Relay
1	8250B-6501	Light/dark module

opcon®
sensing your needs

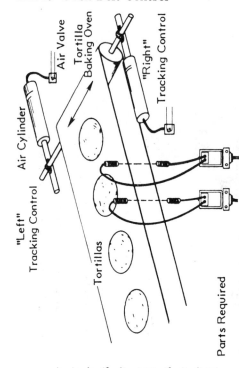

"Left"
Tracking Control

Air Cylinder   Air Valve

Tortilla
Baking Oven

"Right"
Tracking Control

Tortillas

Parts Required

Qty.	Model Number	Description
4	6221A-6501	Thru-beam fiber optic cables with threaded tips
2	1550A-6511	110VAC fiber optic sensors with cables
2	8532A-6501	Relays
2	8250B-6501	Light/dark modules

OVEN BELT CONTROL

Here's an excellent application of fiber optics. (See Figure) An oven used for baking tortillas uses a metal belt to convey the tortillas through the oven. The belt's tendency to move from side to side must be controlled by left and right tracking controls. The controls, of course, must be actuated, and so a pair of fiber optic thru-beam sensors rated for up to 450 degrees Farenheit are placed inside the oven. The control units are mounted outside the oven where it's nice and cool.

Note that the sensor pairs are positioned so that a slight sideways movement of the belt will break the beam and actuate one of the tracking controls. The result is a perfectly aligned belt and better tortillas.

opcon®
SENSING YOUR NEEDS

8.1.20 Sorting Potatoes by Size

SORTING POTATOES
BY SIZE

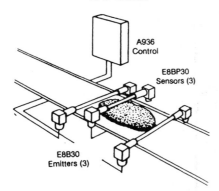

The eye is quicker than the hand, and Autotron Photoelectric Controls are quicker than the eye — and also more discerning. For example, Photoelectric Controls are used to sort potatoes by size on a belt conveyor. The potatoes are run individually, length wise, on the conveyor. Two light beams are established across the conveyor at a distance slightly greater than the desired potato length. A third light beam is established equi-distant between the first two and slightly higher than the desired potato diameter. Potatoes that break all three light beams simultaneously are ejected. Obviously, Photoelectric Controls can be designed to sort almost anything — no matter how unusual its shape may be.

Autotron

8.1.21 Nail Length Inspection

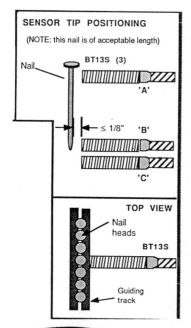

APPLICATION NOTES

the photoelectric specialist   Number AN218

This system is designed to detect whether nails passing on a guided track are too long or too short. When a nail that is too long or too short is detected, a latching output energizes a relay whose contacts can be used to initiate a desired action.

Three SM512LBFO sensors with BT13S fiberoptic assemblies are used in this application. The fiberoptic sensing tips are positioned vertically along the length of the nail as shown in the drawing. When sensor A (the GATE sensor) sees a nail, it sends a GATE signal which then awaits a DATA signal from sensor B. Sensor B outputs a DATA signal only if it sees light (its own light reflected from a nail that is long enough). If this DATA signal is not received (nail is too short) it latches relay BR-2. If sensor C sees reflected light, the nail is too long. It sends a signal which latches the BR-2.

8.1.22 Cookie Motion Detection

COOKIE MOTION DETECTION —————————————————

MODEL #	DESCRIPTION
1550B	Fiber Optic Control
8252A	One-Shot Logic Module (in retrigger mode)
6221A(2)	Fiber Optic Thru-Beam Cable
8532A	Plug-In Relay Output Device

High temperature environments are accommodated by the use of fiber optics. Here conveyor motion in a 450° cookie oven is detected. If the motion stops, the one-shot logic module detects light or dark for too long, and the output device shuts the oven down.

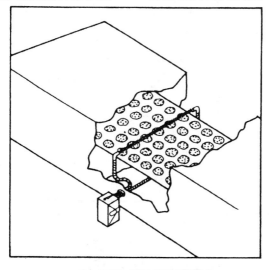

8.1.23 Liquid Level Control

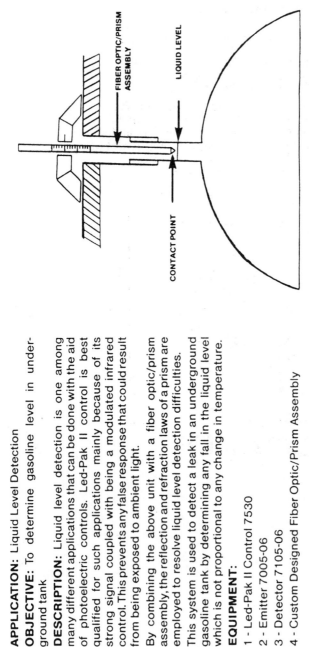

FIBER OPTIC/PRISM ASSEMBLY

LIQUID LEVEL

CONTACT POINT

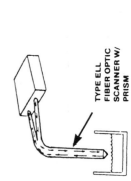

IMMERSION OF PRISM IN LIQUID ACTIVATES LEVEL CONTROL

TYPE ELL FIBER OPTIC SCANNER W/ PRISM

APPLICATION: Liquid Level Detection

OBJECTIVE: To determine gasoline level in underground tank

DESCRIPTION: Liquid level detection is one among many different applications that can be done with the aid of photoelectric controls. Led-Pak II control is best qualified for such applications mainly because of its strong signal coupled with being a modulated infrared control. This prevents any false response that could result from being exposed to ambient light.

By combining the above unit with a fiber optic/prism assembly, the reflection and refraction laws of a prism are employed to resolve liquid level detection difficulties.

This system is used to detect a leak in an underground gasoline tank by determining any fall in the liquid level which is not proportional to any change in temperature.

EQUIPMENT:

1 - Led-Pak II Control 7530
2 - Emitter 7005-06
3 - Detector 7105-06
4 - Custom Designed Fiber Optic/Prism Assembly

Dolan-Jenner

8.1.24 Hopper Fill Level Control

HOPPER FILL LEVEL CONTROL ———————————

MODEL #	DESCRIPTION
8880C-6501	High-Current Control Unit
1381B	Medium-Range Proximity Sensor
8280A	Time Delay Logic Module

The fill level on this machine hopper is controlled by a proximity control with a time delay module. The control must detect the level for a certain time interval before it energizes the shutoff mechanism. This eliminates false "full" indications caused by extraneous material momentarily passing by the control.

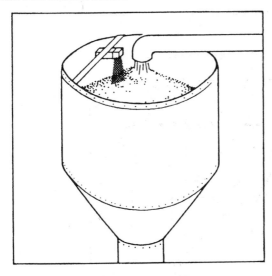

8.1.25 Liquid Clarity Control

LIQUID CLARITY CONTROL ————————————

MODEL #   **DESCRIPTION**
8172A   Analog Control Unit
1173A-100   Moderate-Range Thru-Beam Source
1273A-100   Moderate-Range Thru-Beam Detector

Liquids are monitored in a clear tank with thru-beam controls and an analog control unit. Because the control produces a voltage signal proportional to the amount of detected light, liquid mixtures and densities can be controlled.

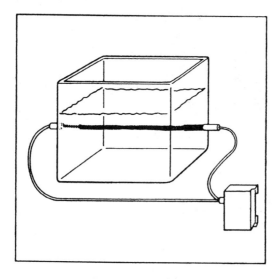

8.1.26 Smoke Detection

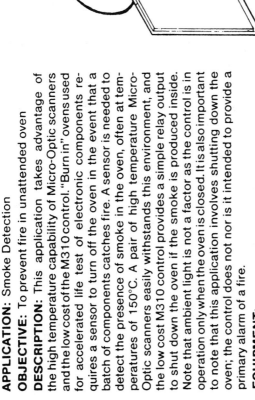

TWO FIBER OPTIC SCANNERS

ELECTRICAL COMPONENTS

CONTROL

150°C

APPLICATION: Smoke Detection

OBJECTIVE: To prevent fire in unattended oven

DESCRIPTION: This application takes advantage of the high temperature capability of Micro-Optic scanners and the low cost of the M310 control. "Burn in" ovens used for accelerated life test of electronic components requires a sensor to turn off the oven in the event that a batch of components catches fire. A sensor is needed to detect the presence of smoke in the oven, often at temperatures of 150°C. A pair of high temperature Micro-Optic scanners easily withstands this environment, and the low cost M310 control provides a simple relay output to shut down the oven if the smoke is produced inside. Note that ambient light is not a factor as the control is in operation only when the oven is closed. It is also important to note that this application involves shutting down the oven; the control does not nor is it intended to provide a primary alarm of a fire.

EQUIPMENT:

1 - Micro-optic control M310

2 - BXT836 Fiber Optic scanners

Dolan-Jenner

8.1.27 Smoke Density Monitor

Smoke Density Monitor

An output of many industrial processes is smoke. Naturally, the density of the smoke must not exceed legal limits, or an entire plant can be shut down. Opcon's 8172A analog control unit provides an inexpensive method for monitoring the clarity of the smoke, and the unit can be used to activate an alarm system before air pollution authorities take note.

A typical smoke-monitoring system uses the following Opcon components:

Qty.	Model No.	Description
1	1173A-100	Light Source
1	1273A-100	Detector
2	6168A-6501	Ball Swivel Mount
1	8172A-6501	Analog Control Unit

See Figure   for a typical installation. Note that the sensors are mounted outside the stack so that the lenses stay clean. This positioning is critical as a deterrent to false triggering; because the application depends on measuring the amount of light blocked by contaminants, the sensor must not be fooled by soot on the lenses.

Overhanging brackets protect the sensor lenses from rain

Sensors transmit light beam through 1" holes in stack. Mounting sensors at a distance from hole keeps them free of soot

0-10 VDC signal is fed to voltage-sensitive relay or chart recorder

1273A-100
Detector

1173A-100
Source

8172A Analog
Control Unit

opcon
sensing your needs

3.1.28 Missing Print Detection

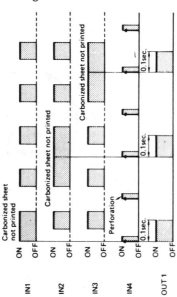

OMRON

Detection of missing print

In the carbonizing and printing process, any carbonized sheet that has print missing is detected and a defect signal is output for a predetermined period.

● Sensors used: Photoelectric switches

IN1: ⎫ Definite reflection type
IN2: ⎬ (LIGHT-ON mode: Model with suffix E1 in type number)
IN3: ⎭

IN4: Separate type (LIGHT-ON mode: Model with suffix E1 in type number)

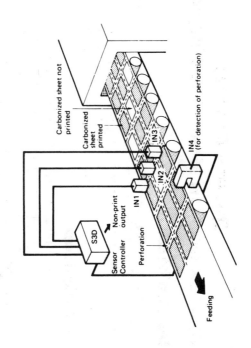

8.1.29 Quality Control: Tile Warp

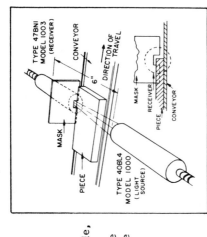

Quality Control
.050" Deflection Means Part Rejection

The Problem

Picture this, a rectangular piece of ceramic tile 2 1/4" long, 7/8" wide, and .094" thick, moving down a conveyor. For this piece to function properly in the customer's equipment, it must lie perfectly flat. If the part is warped .050" or more, it is a reject. Inspection must be made without physical contact with the part. Sounds tough doesn't it?

The Solution

A PHOTOSWITCH 40BL4 Model 1000 Light Source and a 47BN1 Model 1003 Receiver across the conveyor. The light source was placed approximately 4" to 5" from the receiver and was adjusted for the smallest spot possible to be focused so that it ran tangent to the piece. A mask was placed in front of the receiver with an opening 1/16" wide, and .099" high.

As an acceptable piece passes the light beam, .094" of the masked opening is blocked allowing the .050" light beam to reach the receiver. However, if a part warped in excess of .050" passes the light beam, the mask will be completely dark. This energizes the control relay of the 22DJ4 initiating the customer's rejection system.

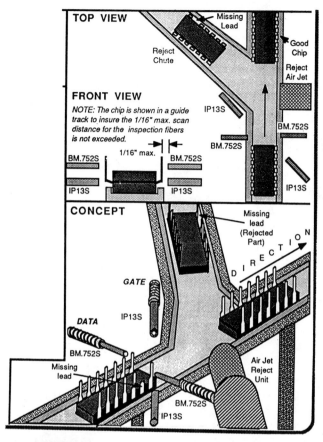

APPLICATION NOTES

the photoelectric specialist

Number AN226

This system is designed to inspect integrated circuit chips to insure that all leads are present before packaging. If a chip is missing one or more of its leads, the system will output a one-shot reject pulse which may be used to control an air jet to eject the bad chip into a reject chute.

The IC chips are conveyed through the inspection area with a gap between parts. The GATE sensor (w/IP 13S fibers) is used in the opposed mode to detect the trailing edge of the chip. The beam is blocked by the body of the chip, and the GATE signal occurs at the dark-to-light transition as the trailing edge passes to check for a valid lead count.

The inspection sensors (w/BM.752S fibers) are used in the proximity mode to count the leads on the chips. A count input occurs at the preset counter when a lead is detected. The BM.752S fiberoptic cables are slightly offset from each other (see top view) to eliminate the chance of optical crosstalk.

8.2 PACKAGING

8.2.1 Label on Glass Bottle Detection

LABEL

LOGIC SCANNER LGX824

7211 TIME DELAY MODULE

NO1 5
C1 4

EQUIPMENT:

1 - Led-Pak II Control 7535
1 - LGX836 Logic Scanner
1 - 7211 Logic Module
1 - Relay 7401

Dolan–Jenner

INPUT SIGNAL

OUTPUT SIGNAL

APPLICATION: Detection of label on glass pill bottle

DESCRIPTION: The LGX836 logic scanner permits the use of a single control to monitor two sense points.

This application involves detection of a label on a amber glass pill bottle. The label covers approximately 240° or 2/3 of the glass on the side as shown in the sketch. The small bottles pass the sensing point oriented at random with respect to the scanner position. The contrast between the label and the amber bottle was found to be very reliable using the Led-Pak II control. In order to reliably detect the presence of the label, two sense points 180° apart are required. This is an application where the LGX836 logic scanner can be employed to advantage. Using the two sensor ends, either of which will detect the label, the use of multiple emitters and detectors is avoided. This ensures more reliable operation by requiring fewer electrical components, and simplifies the interconnect wiring.

There is one case which might give a false signal: if a bottle with a label passes by, and the 1/3 part which is unlabeled comes in a horizontal position with the scanners, then the beam will be broken for a period of time less than that of an unlabeled bottle.

To avoid the short pulse from giving a false signal, a time delay module (7211) is used and the ON DELAY, T2 is adjusted so that a short pulse caused by the case above can be ignored.

8.2.2 Cap and Bottle Presence Detection

CAP & BOTTLE PRESENCE

Application: Make sure both cap and bottle are present in high-speed bottling plant.

Opcon products used: 8700 Series

Quantity	Model #	Description
2	8772A	Control unit with time delay/one shot
2	1173A-300	Source
2	1273A-300	Detector
2	8530A	Relay
2	8905A	Socket
2	8907A	Socket
2	8133A	Power supply

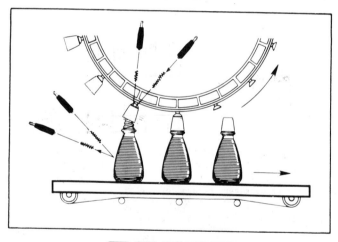

8.2.3 Cap Orientation Detection

CAP ORIENTATION DETECTION

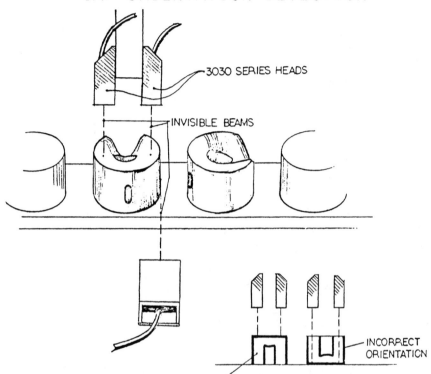

- 3030 SERIES HEADS
- INVISIBLE BEAMS
- INCORRECT ORIENTATION
- CORRECT ORIENTATION

A triple sensor 3030-series is
being used to monitor aerosol can
tops as they pass on a conveyor.
One sensor activates the pair of sensors looking
down on the caps. A pair is necessary because of the
depression in the cap which maybe randomly oriented.
As long as one sensor sees a surface of the cap, the cap
is properly oriented. If both sensors do not see the
surface of the cap, the cap is upside down on the
conveyor.

FOR DETECTING
AND REJECTION OF
UPSIDE-DOWN CAPS.

SCIENTIFIC TECHNOLOGY INCORPORATED
ELECTRO-OPTIC DEVICES, CONTROLS AND SYSTEMS

Reprinted with permission from Scientific Technology Incorporated.

8.2.4 Bottle Fill Level and Cap Detection

Detection of fill level and bottle cap

On a bottling machine, the S3D sensor controller and three sensors detect the presence of a bottle cap and check the fill level of each bottle. Three separate failure signals from the S3D controller indicate the bottle cap is missing, the fill level is insufficient, or both cap is missing and fill level is insufficient. The failure output signals stay on for a predetermined period.

As the conveyor brings the bottles to the checkpoint, the first sensor (IN1) checks for a cap. The material of the cap determines the type of sensor used: A proximity sensor for metallic caps, a definite reflection-type photoelectric sensor for plastic caps.

Fill level is checked by a beam-break photoelectric sensor (IN2). Another one (IN3) detects a bottle on the conveyor and synchronizes the inspection.

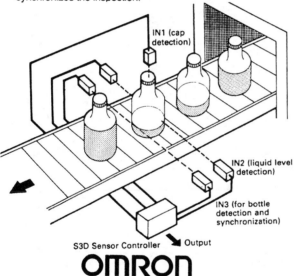

IN1 (cap detection)

IN2 (liquid level detection)

IN3 (for bottle detection and synchronization)

S3D Sensor Controller   Output

OMRON

8.2.5 Reversed Cap Detection

Reversed cap detection

When a reversed cap is detected among the caps on the conveyor feeding the automatic capping machine, a defect signal is output for a predetermined period.

- Sensors used: Photoelectric switches
  - IN1: Definite reflection type (LIGHT-ON mode: Model with suffix E1 in type number)
  - IN2: Separate type (DARK-ON mode: Model with suffix E2 in type number)

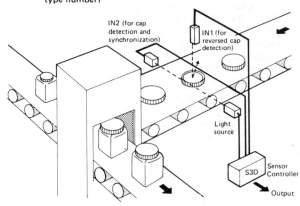

- **Operation chart**

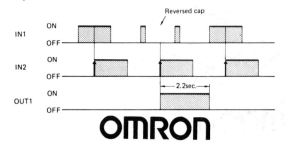

OMRON

8.2.6 Liquid Level Detection

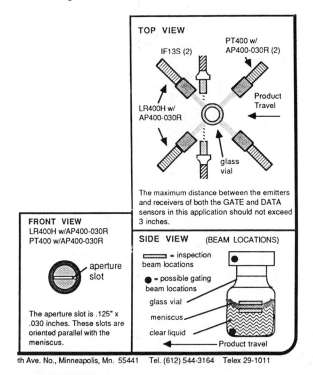

FRONT VIEW
LR400H w/AP400-030R
PT400 w/AP400-030R

aperture slot

The aperture slot is .125" x .030 inches. These slots are oriented parallel with the meniscus.

th Ave. No., Minneapolis, Mn. 55441   Tel. (612) 544-3164   Telex 29-1011

BANNER **APPLICATION NOTES**

the photoelectric specialist

Number AN203

This system inspects for the level of a clear liquid in a clear glass vial, and rejects the vial if the level of the liquid is either above or below acceptable limits.

The model SM512LBFO sensor (with IF13S fiberoptic assemblies) is used to create the interrogate (GATE) beam for the system. The Side View (below) shows two possible locations for the GATE beam. When a bottle moving along the conveyor interrupts this GATE beam, the two LR400H/PT400 sensor pairs (see Top View, below) inspect the bottle for the level of the liquid. These inspection (DATA) sensor pairs use MB3-4 amplifiers in "dark operate" mode, and are apertured as shown to restrict their field of view (see Front View). One pair is set to "look" slightly below the acceptable liquid level, and second pair to "look" slightly above the acceptable level. These sensors look for the edge of the meniscus (the curved upper surface) of the liquid in the vial.

8.2.7 Liquid Fill Level Detection

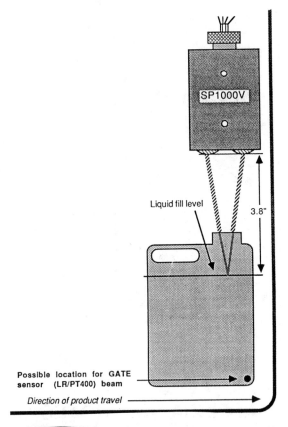

Liquid fill level

3.8"

SP1000V

Possible location for GATE sensor (LR/PT400) beam

Direction of product travel

APPLICATION NOTES

the photoelectric specialist

Number AN217

This system checks opaque plastic bottles to determine if they have been filled to the proper level prior to capping. As the bottles move along a conveyor, they break the GATE beam of the opposed mode LR/PT 400 sensor pair, which is set up to be looking across the conveyor in the path of the bottles. The SP 1000V convergent-beam inspection sensor is positioned so that it is looking downward (through the open bottle neck) at the upper surface of the liquid the moment the GATE beam is broken. If the bottle is filled to the proper level, the SP 1000V will "see" it's own light reflected from the liquid's surface and output a DATA signal. If the bottle is underfilled, no reflected light is seen, no DATA signal is sent. The absence of a DATA signal with the GATE is used to produce a one-shot pulse. This pulse can be used to activate a bottle rejection mechanism, sound an alarm, etc.

8.2.8 Wine Bottle Missing Cork Inspection

APPLICATION: Inspection of wine bottles on a moving conveyor for missing corks.

OBJECTIVE: To shut down the machine if 5 consecutive corks are missing.

DESCRIPTION: Two 7530 controls are used for this application. Control #1 provides a clock pulse to the shift register 7217 while control #2 provides an input signal every time it sees a cork. The 7213 one-shot module on control provides a clock pulse shoter than the input pulse. A third control (LED-PAK IV with 2300 I/O Module) is used to reset the shift register if less than 5 consecutive corks are missing.

EQUIPMENT:

2 - 7530 Controls

1 - 7213 One-Shot Module

1 - 7217 Shift Register Module

1 - LED-PAK IV with 2010 Electronic Module and 2300 I/O Module

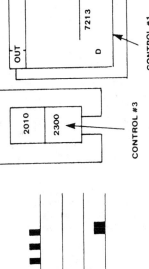

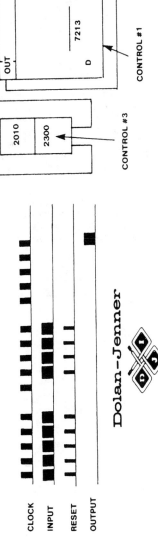

Dolan-Jenner

8.2.9 Container Spacing Adjustment

SPACE ADJUSTMENT
WITH DUAL TIME DELAY CONTROL

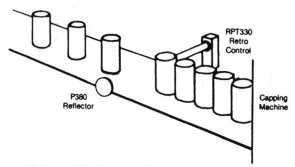

When applied to bottling lines, Photoelectric Timing Controls with dual time delay controls help operate bottle cappers more efficiently. Several bottles touching each other are required to operate the bottle capper, but bottles often come on the conveyor with spaces between them. The dual timing control helps maintain uniformity in spacing by timing out light and also timing out dark. When bottles break the light beam continuously through the dark timing period, the control relay energizes and starts the capping mechanism. When spaces between bottles keep the light beam restored through the light timing period, the control de-energizes and stops the bottle capper.

8.2.10 Fallen Bottle Control

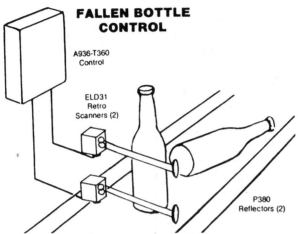

FALLEN BOTTLE CONTROL

A936-T360
Control

ELD31
Retro
Scanners (2)

P380
Reflectors (2)

After the bottle capping operation is completed, bottles sometimes fall over on the conveyor. This will eventually interfere with the packing operation, and the conveyor must be stopped. The Photoelectric Control shown above is equipped with two retro scanners mounted on the side of the conveyor. The Photoelectric Control stops the conveyor when the lower light beam is broken and other light beam is not broken. Each upright bottle breaks both light beams simultaneously, and each space between upright bottles restores both light beams. When a fallen bottle interrupts only the lower light beam, the Photoelectric Control operates to stop the conveyor immediately.

8.2.11 Aerosol Can Inspection

Inspection

The MCS-651 two part, thru-beam scanner is the ideal solution to detect the presence/absence of a spray nozzle on these aerosol cans. The modulated LED light source is virtually immune to false signals from ambient light. Size of the scanner is 1" square by ¼" thick.

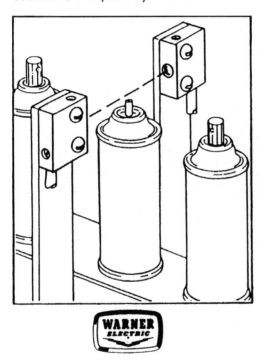

8.2.12 Carton Sealing

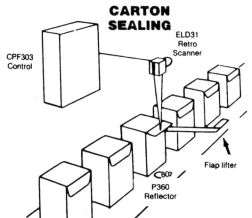

Photoelectric quality control assures that all cartons passing through the control point on this conveyor are properly sealed. Improper seals on cartons are easily spotted by the Photoelectric Scanner mounted on top of the conveyor. Improperly sealed cartons are automatically ejected from the conveyor. After cartons are sealed, the conveyor takes them through the "flap lifter." As each carton passes, flaps not securely glued are caught by the flap lifter and lifted to break the Photoelectric Control light beam. An ejecting device then removes the improperly sealed carton. Where cartons have flaps on both sides, two flap lifters are used, either of which will operate the ejector.

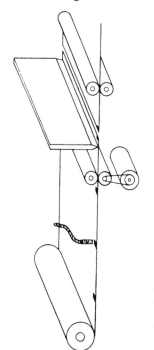

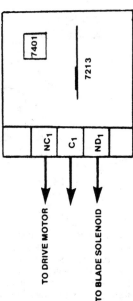

TO DRIVE MOTOR

TO BLADE SOLENOID

APPLICATION: Registration Mark Detection

OBJECTIVE: Precise Label Cutting

DESCRIPTION: Fiber optic controls can be used to operate a solenoid actuated blade in automatic cut off systems. The original products come in a form of webs or rolls and must be processed before reaching a final stage. The application is to precisely cut different sizes of labels.

The Micro-Optic II control (8230) is used because registration marks may vary in color and reflectivity. The application involves detecting the registration mark, stopping drive motor and activating cutting blade simultaneously.

Using a one-shot logic module 7213 and also using both normally open and normally closed relay outputs this can be accomplished reliably.

EQUIPMENT:

1 - Micro-Optic II Control 8230
1 - 7060-06 Emitter
1 - 7160-06 Detector
1 - One Shot Module 7213
1 - DPDT Relay 7401

Dolan-Jenner

8.2.14 Tube Orientation

ORIENTING TUBES
OR OTHER ROUND CONTAINERS

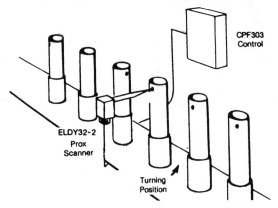

CPF303
Control

ELDY32-2
Prox
Scanner

Turning
Position

Uniformly orienting tubes or other round containers for a filling, capping and/or packing operation requires the speed and precision of a Photoelectric Control. Uniform orientation can be accomplished very easily with a scanner mounted on the side of a conveyor. The illustration above shows toothpaste tubes with the cap end down, bottom up and open. A register mark near the tube bottom is used for orientation. If the Photoelectric Control fails to see the mark, it engages the spinning clutch until the mark turns into the correct position and then disengages the clutch. (The conveyor stops when each tube moves into spinning position).

8.2.15 Label Position Control

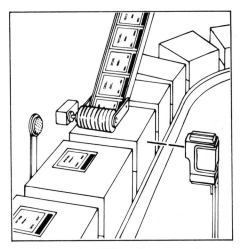

Position Control
With its internal timing function, the
MCS-500 series can detect box and
activate a delayed one shot signal to
have labels applied.

8.2.16 Shiny Container Detection

SHINY MATERIAL DETECTION

Shiny materials: Use the retro mode with caution when sensing shiny materials. Even though the optics of good retro scanners are designed with great care to minimize "proxing", a shiny surface may render a retro scanner "blind" to its target. If a retro scanner must be used, scan across the shiny material at a skew angle to the material's surface.

If possible, use a visible retro scanner with polarizing filter and a corner cube reflector to eliminate the chance of proxing.

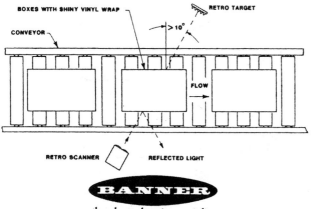

the photoelectric specialist

8.2.17 Indexing: Shrink Wrap

INDEXING — SHRINK WRAP

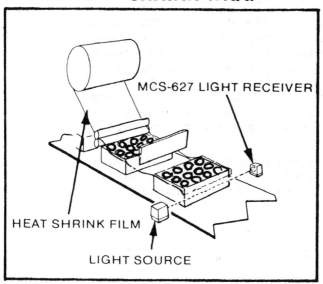

MCS-627 LIGHT RECEIVER

HEAT SHRINK FILM

LIGHT SOURCE

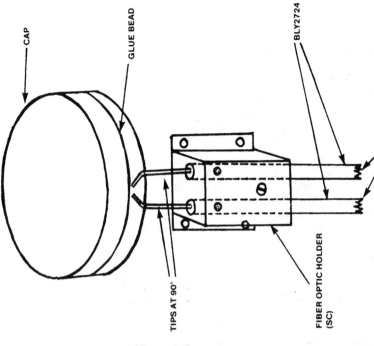

CAP

GLUE BEAD

BLY2724

TO MICRO-OPTIC II CONTROL. (EMITTER & DETECTOR)

TIPS AT 90°

FIBER OPTIC HOLDER (SC)

Dolan–Jenner

APPLICATION: Detection of Bead of Glue

OBJECTIVE: To ensure uniformity of gluing operation

DESCRIPTION: This is an application where a bead of glue needs to be detected in order to assure good quality on the finished products.

A reflective scanner E-Type cannot be used in this case due to low contrast between the glue and the background, white on white. Therefore, another approach is used by positioning 2 BLY2724 scanners at a right angle with respect to each other, with sensitivity of the control adjusted so that as long as the fiber optic scanner is seeing the glue bead it will deflect the beam away from the receiver indicating the presence of a good part. Otherwise, the scanner will see the white background and the beam will be reflected to the receiver causing the control to activate a rejection mechanism or sound an alarm.

EQUIPMENT:

1 - Micro Optic II Control 8230

1 - 7060-06 Emitter

1 - 7160-06 Detector

2 - BLY2724 Fiber Optic Scanners

1 - Fiber Optic Holder SC

1 - DPDT Relay 7401

8.3 MATERIAL HANDLING

8.3.1 Conveyor Jam Detection

JAM DETECTION ————————————————————

MODEL #	DESCRIPTION
1450B	Reflex Control
8251A	Time Delay Logic Module
8532A	Plug-In Relay Output Device
6200A-3	Retroreflector

A reflex control with a time delay module set for "delay dark" ignores momentary beam breaks. If the beam is blocked longer than the delay period, the output energizes to sound an alarm or stop the conveyor.

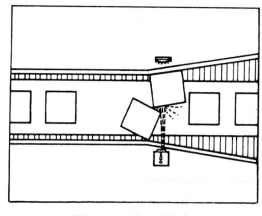

8.3.2 Container Size Sorting: Two Sensors

Case Sorting by Size

Two MCS-500 Photoscanners detect box size and outputs signal to divert boxes to the appropriate work station.

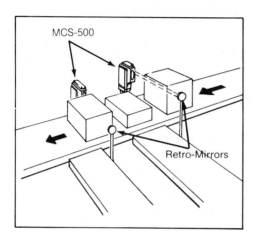

8.3.3 Container Size Sorting: One Sensor

Application: Sort containers by size on a single conveyor using one sensing location.

Opcon products used: Sensing: by 70 Series heads; Decision Logic: by 8700 Series modules; Shift Logic: by 8730 and/or 8731 Shift Registers. The specific model numbers will vary depending on the requirements of each application. Please contact the Opcon application engineering department for product recommendations.

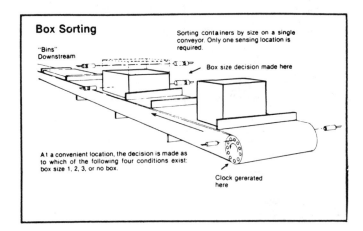

8.3.4 Container Sorting by Retro Tape

SORTING CARTONS
BY RETRO TAPE MARKS

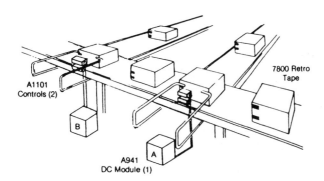

No labor is required in this totally automated sorting operation. No slow downs or sorting errors occur as long as Photoelectric Controls do the job of identifying and sorting cartons by size, content, destination or other criterion. Cartons have retro tape marks in different arrays. Cartons with one array are shoved onto the proper side conveyor by Control A which operates a pushing device. Cartons with another array pass through Control A and are shoved onto another side conveyor by Control B. Cartons with all together different arrays stay on the main conveyor. This is one of many possible sorting and distribution operations using Autotron Photoelectric Controls.

8.3.5 Shift Register Delayed Ejection

SHIFT REGISTER CONTROL

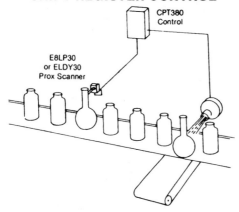

CPT380
Control

E8LP30
or ELDY30
Prox Scanner

    Sorting containers of different heights and shapes
on a conveyor is done automatically by Photoelectric
Controls without interrupting the movement of the conveyor. This application requires one scanner which
operates a Photoelectric Control, as shown above. The
light beam is sufficiently elevated so that only tall bottles are detected. When the sensor detects a tall bottle,
the Control relay energizes, when the bottle reaches
the ejection mechanism, causing the removal of the tall
bottle. The Control has the ability to "remember" the
tall bottle until it reaches the ejection point, while all
other bottles below the height of the light beam continue along the conveyor.

8.3.6 Carton Counting

BOX COUNTING

Application: Detect and count cartons anywhere on four foot belt in any orientation; provide total count at end of shift.

Model#	Description
1410B	Control unit
6200A-3	3 inch reflector
8220A	Totalizing counter

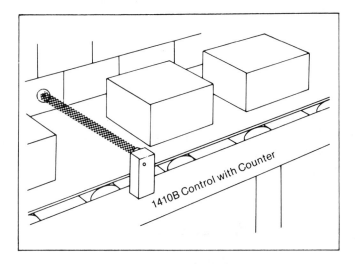

1410B Control with Counter

8.3.7 Batch Counting

BATCH COUNTING AND DIVERTING ─────────────

MODEL #	DESCRIPTION
1451B	Polarized Reflex Sensor
8254A	Divider Logic Module
6200A·3	Retroreflector
8532A	Plug·in Relay Output Device

Cans on a conveyor are diverted to two other con-veyors controlled by a polarized reflex control with a divider module. Items can be diverted in groups of 2, 6, 12, or 24. A polarized control is used, so shiny cans do not falsely trigger the control.

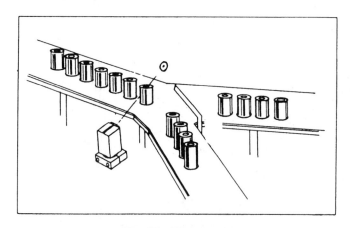

8.3.8 Pallet Load Wrapping Control

**PALLET LOAD
WRAPPING CONTROL**

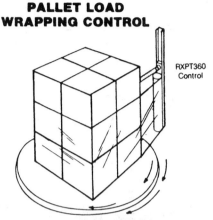

RXPT360
Control

The pallet load in the above application is being wrapped in a protective plastic sheet. The roll of wrapping material on the post facing the pallet load moves up as the sheet is wound around the pallet load. A Photoelectric Control mounted just above the top of the roll monitors the height of the pallet load. As long as the pallet load reflects the light beam back to the sensor (in the same unit as the light source), the roll continues to move up the post. When the light beam clears the top of the pallet load, the reflection is lost and the Control stops the wrapping operation. A brief time delay prevents false stops due to momentary loss of reflection from highly irregular shaped loads.

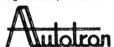

8.3.9 Palletizer Off-Load Control

PALLETIZER OFF-LOAD CONTROL

In the accompanying illustration, an off-load conveyor is divided into two independently driven sections. The palletizer outfeed is also driven separately. Three infrared light beams control the conveyor and palletizer out-feed. When a pallet breaks beam #1, conveyor section A is shut down, thus preventing the pallet from being pushed off the end. As a second pallet moves down the line it breaks beam #2. The DPDT relays used in the control units are interconnected so that if both beam #1 and beam #2 are blocked, conveyor section B stops, a third pallet attempting to enter the off-load line will break beam #3. If both beam #3 and beam #2 are broken, the palletizer out-feed is shut down. Removing the pallet blocking beam #1 will restart conveyor section A which restarts section B, etc.

Equipment required:
3 ea Reflex Control Units
    Models 1410 or 1420 may be selected
    Model 1411B with plastic case may also be suitable
    Models 8170A or 8171A control units with a 1470 reflex head

3 ea DPDT Relays
    Select from models 8526, 8529, or 8530 to suit control unit used

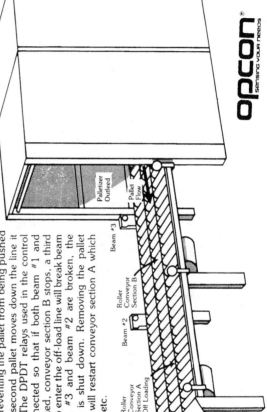

Palletizer Outfeed
Pallet Flow
Beam #3
Roller Conveyor Section B
Beam #2
Roller Conveyor Section A Off Loading
Beam #1

opcon
SENSING YOUR NEEDS

8.3.10 Stack Height Control

STACK HEIGHT CONTROL ——————————————

MODEL #	DESCRIPTION
1155A	Thru-Beam Source
1250B	Thru-Beam Detector
8250B	Light/Dark Logic Module
8532A	Plug-In Relay Output Device

A set of thru-beam controls determine the height of a scissor lift. For example, when the control is set for "dark-to-light" energize, the lift rises after a layer has been removed and stops when the next layer breaks the beam again.

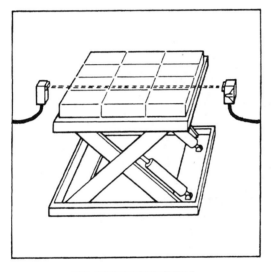

opcon®
sensing your needs

8.3.11 Log Diverter

LOG DIVERTER ────────────────────────────────

MODEL #	DESCRIPTION
8890A	Power Base
1391A	Long-Range Proximity Sensor
8252A	One-Shot Logic Module

The log is detected by a long-range proximity control with a one-shot logic module. The hydraulic system is energized and the log is pushed onto a lateral conveyor. The logic module allows for control of output pulse timing and duration.

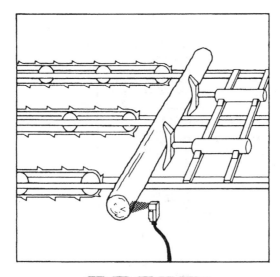

opcon®
SENSING YOUR NEEDS

8.4 MACHINE TOOL

8.4.1 Bent/Broken Drill Bit Detection

BENT/BROKEN DRILL BIT DETECTION

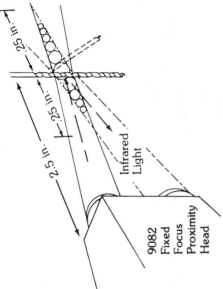

2.5 in.

.25 in.

.25 in.

Infrared
Light

9082
Fixed
Focus
Proximity
Head

Automated machinery has become a standard part of the manufacturing scene. The complexity and speed of automated processors, however, has made it difficult for the human operator to adequately monitor the functional status of the tools used. The smaller the tool, the more critical and difficult the monitoring task becomes. Small drill bits, for example, break or bend easily and are best checked after each drilling operation. Either proximity or through-beam detection methods are suitable for detecting or monitoring small objects.

The proximity method is relatively straightforward. Opcon's 9082 fixed-focus proximity head is the key to reliable small parts detection. The 9082 produces an intensely focused detection region about .125 inches wide, extending .25 inches fore and aft of the focal point. The focal point is about 2.5 inches from the 9082. The 9082 is capable of detecting drill bits less than .037 of an inch in diameter. The minimum detectable diameter will vary with conditions in the working environment. Although the focal range is about 2.5 inches, the actual detection region extends ± .25 inches. It may be necessary to experiment within the range limits for best results.

Infrared light from the source strikes the drill bit and some of it is reflected toward the detector (detector see light). If the drill bit is bent or broken, little or no light will be reflected (dectector see dark). To prevent multiple outputs when the bit is bent, a time delay or one-shot module may be necessary.

Through-beam detection of small object requires the use of apertures for detection of objects less than the effective beam diameter. On the plus side, however, through-beam offers relatively high power to see through contaminants such as cooling or lubricating sprays, provided the apertures do not unduly restrict the beam. The through-beam approach is also comparatively easy to squeeze into tight places, although wiring the two separate heads is somewhat inconvenient.

opcon®

SENSING YOUR NEEDS

8.4.2 Broken Drill Bit Detection

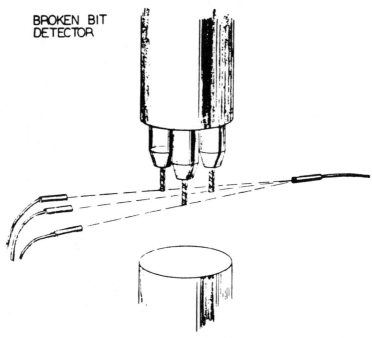

STI sensors are used for detecting broken bits or taps on automatic machines. In the illustration shown, a single transmitter is aligned so that the tips of the bits break the beam to 3 receivers. If a bit breaks, a beam is made, and the machine shuts down.

STi ₁ₘ **SCIENTIFIC TECHNOLOGY INCORPORATED**
ELECTRO-OPTIC DEVICES, CONTROLS AND SYSTEMS

Reprinted with permission from Scientific Technology Incorporated.

8.4.3 Broken Drill Bit Detection: Fiber Optics

Broken Drill Bit Detector

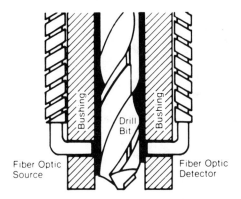

Fiber optics can be used in tight
spaces to detect broken drill bits.
If the drill is broken or missing,
light energy will freely pass from
the source fiber to the detector
fiber. If the drill is present, the
light will be blocked.

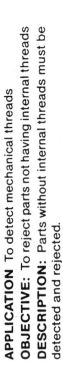

APPLICATION To detect mechanical threads

OBJECTIVE: To reject parts not having internal threads

DESCRIPTION: Parts without internal threads must be detected and rejected.

The approach is to reflect the light beam from the smooth surface of the part at a 90° angle when an unthreaded part is passing the inspection station; this will activate the detector.

The output of the control is used to trigger a rejection mechanism that will reject the part.

A 7213 logic module can be used to preset the on-time of the rejection device if required. If a component is properly threaded the reflected light will not reach the detector.

EQUIPMENT:

1 - Micro-Optic II Control 8230

1 - 7060-06 Emitter

1 - 7150-06 Detector

2 - BT624 Fiber Optic

2 - LH501 Lenses

2 - FOA-1 Adapters

1 - DPDT Relay 7401

Dolan-Jenner

8.4.5 Part Detection: Chute

Chute/Part Detection

In vibratory feeders, for example, vibration is often
a problem for many photoelectrics. Fiber optic
cables are the answer. They fit in tight spots,
there is no filament to break and all electronics
are remotely located in the control housing.

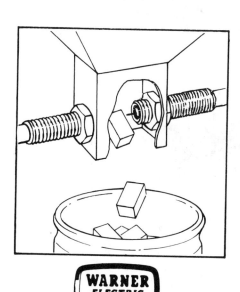

8.4.6 Small Parts Detection

SMALL PARTS DETECTION ————————————

MODEL #	DESCRIPTION
8171B	High Sensitivity Control Unit
8215A	Low Contrast Logic Module
1471A	Reflex Curtain of Light Sensor
6210A	Retroreflector
8572A	Triac Output Relay

Small objects moving through a "curtain of light" are counted by detecting a change in reflected light. A low-contrast logic module inside the control unit responds to slight but abrupt signal variations while ignoring slow changes such as dust buildup.

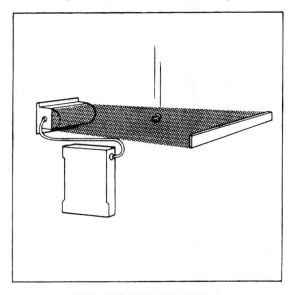

8.4.7 Part Detection: Robotic

Robotics-Part Detection

Fiber optic heads are the right answer for this popular robotic application. Straight type or right angle heads have been designed for tight, space restricted areas and harsh environments involving shock, vibration and heat.

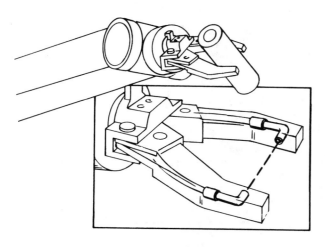

8.4.8 Die Casting Machine Protection

AUTOMATIC PROTECTION
FOR DIE CASTING MACHINES

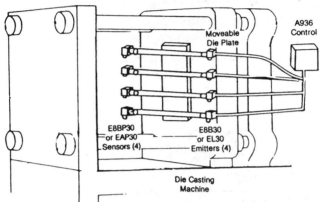

This die casting machine has automatic, built-in protection against castings that stick in the die. The emitters and sensors of the photoelectric control are mounted on opposite sides of the moveable die plate, as shown in the above sketch. With the moveable die plate in the extreme open position, the light beams strike the sensors, causing the relay to recycle the timing control of the press for another stroke. When a casting sticks in the die, it interrupts one or more light beams and stops the machine. The photoelectric control could also sound an alarm when a casting sticks. This is one of many applications that could prevent damage to machines and reduce waste.

8.4.9 Part Detection: Assembly

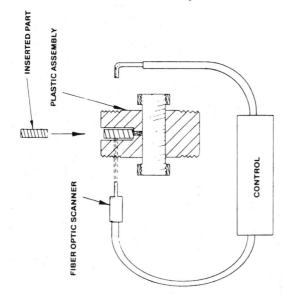

Dolan–Jenner

APPLICATION: Missing Part Detection

OBJECTIVE: Verification of presence of a part within another part

DESCRIPTION: The powerful LED source of the Led-Pak II control provides the ability to "see through" many plastic materials. Using this ability, a control can be set up to scan through an assembly and look for missing parts. In the example shown a small precision spring is inserted into a valve body. The assembly of the complete valve is accomplished on an automated seven station rotary table. One of the problems in automation is to devise a means of verifying that each part in the assembly is present before proceeding. The precision spring presents a particularly difficult sensing challenge. It can not be detected reflectively from the top as the spring outside diameter was too close to the inside diameter of the valve opening. However, the strong infrared beam of the Led-Pak II can penetrate the opaque plastic material of the valve body. A further problem is presented by the very fine cross section of the spring coils. With emitter intensity sufficiently high to pass through the part, the wire of the spring coil could not block the through scan beam. The solution is to use the very small BLY2724 (0.027" (7mm) bundle diameter) scanner on the 7105 receiver end.

EQUIPMENT:

1 - Led-Pak II control 7535
1 - B836 Micro-Optic scanner
1 - BLY2724 Fiber Optic scanner
2 - DPDT Relay 7401

8.4.10 Part Orientation

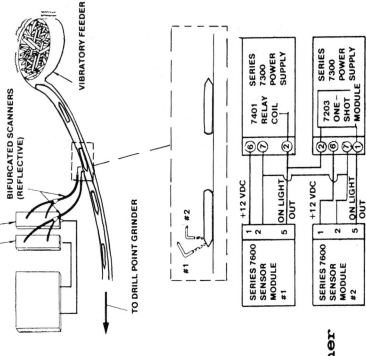

VIBRATORY FEEDER

BIFURCATED SCANNERS (REFLECTIVE)

CONTROLS

TO DRILL POINT GRINDER

#1   #2

SERIES 7600 SENSOR MODULE #1				+12 VDC		7401 RELAY COIL	SERIES 7300 POWER SUPPLY
1	2		5	ON LIGHT OUT		⑥ ⑦ ②	

SERIES 7600 SENSOR MODULE #2				+12 VDC		7203 ONE-SHOT MODULE	SERIES 7300 POWER SUPPLY
1	2		5	ON LIGHT OUT		② ⑥ ⑦ ①	

NOTE: ONE-SHOT MODULE IN DARK OPERATE POSITION

Dolan-Jenner

APPLICATION: Part Orientation

OBJECTIVE: To detect parts improperly oriented

DESCRIPTION: The small size of our ELY2724 scanners makes possible a degree of resolution not realizable with traditional photoelectric controls. Using two of these scanners with Led-Pak I sensor modules and power supplies, a logic condition is set up capable of sorting drill bit blanks. The blanks are fed from a vibratory bowl feeder to the drill point grinder that grinds the flutes into the blank. The blanks, however, have a preground pointed tip that must be presented to the grinder. Sorting the blanks is accomplished using the two controls shown in the diagram. Scanner and sensor module marked #2 reflectively detect the cylindrical edge of the drill blank and allow the one-shot modules output to turn off briefly. The sensor module marked #1 is wired directly to the one-shot output and will hold the output relay "on" if the preground point of the blank is present. If the blunt end is presented first, the output relay will de-energize and activate a solenoid to eject the blank back into the bowl.

EQUIPMENT:

2 - Led-Pak I sensor modules 7611
2 - Power supply modules 7310-00
1 - One-shot logic module 7203
1 - DPDT relay 7401
2 - Fiber Optic scanners ELY2724

8.4.11 Control of Sequential Operations

Control Sequential Operations From One or More Detection Points

Consider the task of punching holes in the leading edge of a sheet of metal which is riding flat on a conveyor surface. An Opcon 1370A proximity detector senses the approach of the sheet of metal to the punching station. The signal from the 1370 causes the 8771 demodulator module to give an output signal to 8713 one-shot module #1.

This module provides a delay to give the sheet of metal time to come to rest against two pins which block its path on the conveyor. When the preset time has elapsed, one shot module #1 triggers one-shot modules #2 and #3 simultaneously.

One-shot module #2 activates Opcon model 8572 triac module #1 for a preset period of time. This causes a crowder to push to the sheet of metal over against a rail on the side of the conveyor. The sheet of metal is then properly positioned for punching.

One-shot module #3 provides a delay which gives the crowder time to do its job and then get out of the way. When the preset time has elapsed, one-shot module #3 triggers one-shot modules #4 and #5.

One-shot module #4 activates triac module #3 for a preset period of time. This causes the punch to punch holes in the sheet of metal.

One-shot module #5 provides a delay which gives the punch time to do its job and get out of the way. When the preset time has elapsed, one-shot module #4 triggers one-shot module #6.

One-shot module #6 activates triac #2 for a preset period of time. This causes the pins which blocked the progress of the sheet on the conveyor to retract long enough for the sheet to proceed. The process is now complete.

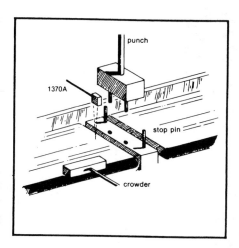

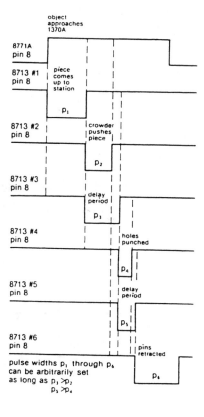

8.5 PEOPLE AND VEHICLE DETECTION

8.5.1 Tollbooth Control

TOLLBOOTH CONTROL ————————————————

Application: Traffic control, make certain all vehicles pay toll before raising gate; lower gate instant that end of paid vehicle passes sensor to eliminate toll cheating. Outdoors, all weather, 24 hour operation.

Opcon products used: Self Contained Series.

Model #	Description
1420B	Control unit reflex sensing, 115 VAC, sealed metal housing.
6200A-3	Reflector
8530A	Relay

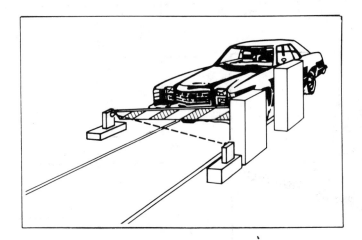

8.5.2 Vehicle Over-Height Detection

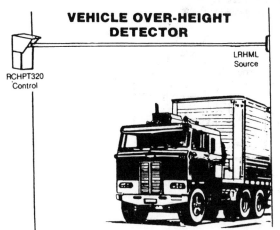

Strategically located Photoelectric Controls in the docking areas of truck terminals and warehouses can be a major factor in preventing damage and accidents. The above illustration shows a vehicle over-height detector used to warn drivers of an approaching low-clearance area. When the light beam is broken by the trailer (or cab) an alarm operates to give advance warning of an impending collision. The Photoelectric Control has high immunity to ambient light, even sunlight. Anti-fog lens heaters and hoods prevent malfunction due to lens fogging or accumulation of rain, ice or snow. The Control provides a one-shot relay operation to avoid continuous alarm when a vehicle stops in the light beam.

8.5.3 Vehicle Detection: Drive-Up Window

VEHICLE DETECTOR
AT DRIVE-UP WINDOWS

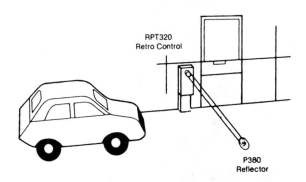

RPT320
Retro Control

P380
Reflector

Many banks, restaurants and other businesses use Photoelectric Controls to detect vehicles at drive-up windows. This allows employees working in other areas to service drive-up windows without making customers wait. When it reaches the drive-up window, the vehicle breaks the light beam back to the sensor located in the same unit as the light source, and the control sounds a buzzer. An automatic timing period on the Control shuts off the buzzer if the car remains parked. When the vehicle leaves, the timing period resets itself for the next vehicle.

8.5.4 Customer Detection: Bank Teller Window

CUSTOMER DETECTION
AT BANK TELLER WINDOWS

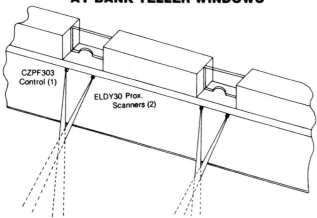

CZPF303
Control (1)

ELDY30 Prox.
Scanners (2)

Waiting at the teller window when the teller does not know you are there can be disturbing for customers. Photoelectric controls can increase efficiency of teller operations and improve customer relations. In the above sketch, dual photoelectric scanners are installed at each window. When a customer is at the window, the invisible (infrared) beams detect him by diffuse reflection. This signals the teller to come to the window. Each pair of scanners are small enough to be mounted inconspicuously under the counter. Each pair is connected to a single control so that detection by either scanner is sufficient. This allows for wider area coverage.

8.5.5 Long-Range Perimeter Security

LONG-RANGE
LED DETECTION SYSTEM

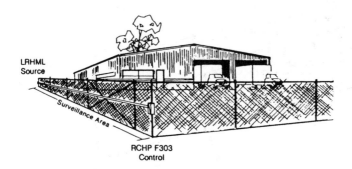

LRHML
Source

Surveillance Area

RCHP F303
Control

The Autotron Long Range LED Detection System shown in the above application is ideal for improving security for warehouses, garages, outdoor fenced-in storage areas or indoor restricted areas. The LED light beam spans a maximum distance of 200 feet and is completely immune to ambient light, even sunlight. The light source projects an infrared light beam which no intruder can detect. When the invisible light beam is blocked, the Control could operate an alarm or lights. The above application shows the light source and sensor mounted on posts at opposite ends of a chain link fence. If a higher or lower light beam is required, the light source and sensor can be adjusted easily.

8.5.6 Automatic Car Wash Control

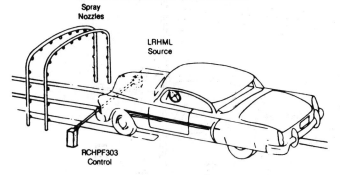

**AUTOMATIC
CAR-WASHING CONTROL**

Spray
Nozzles

LRHML
Source

RCHPF303
Control

Photoelectric controls operate dependably even under extremely adverse environmental conditions. Every operation of an automatic car-washing facility can be regulated precisely by photoelectric controls as the car is pulled through the facility by a continuous low cable. For example, when the car interrupts the light beam, the control relay starts the spraying equipment. When the light beam is restored, the spraying stops. The powerful beam of the light source assures reliable operation despite spray and mist. The anti-fog lens heater (H) prevents lens fogging from the humid air. Photoelectric controls not only assure a satisfactory car wash every time, they also help control the cost of utilities.

8.5.7 Automatic Door Control

**AUTOMATIC
DOOR CONTROL**

RCPF303
Controls (2)

LRML
Sources (2)

The advantages of the Photoelectric door control shown above are : (a) the door is self-closing and self-opening, and (b) the door may be opened or closed automatically from the inside or outside. The door opens when the first light beam is interrupted and closed when both light beams are restored. Traffic entering and exiting is greatly expedited. If street traffic is heavy in the area where vehicles enter, Photoelectric door controls can make entering and exiting much safer, since vehicles are stopped only momentarily while the door opens. Increased security is another advantage of photoelectric door controls. When the Control is turned off, the door is locked until the Control is turned on.

8.5.8 Automatic Door Control

Automatic Door Control

Reflective/proximity or retroreflective scanners
with retro mirrors sense person's approach.
The Photoscanners' multi-function timer is set
to hold the door open for a pre-determined time.

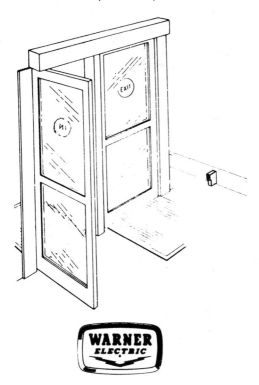

8.5.9 Collision Avoidance

COLLISION AVOIDANCE

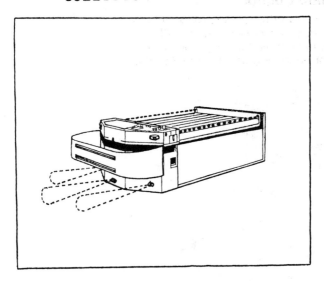

Opcon AGV collision avoidance photoelectric
sensors provide warning up to eight feet in
front of the vehicle to provide additional
stopping distance not available from mechanical
bumpers when guidepath obstructions are
encountered.

8.5.10 Photoelectric Race Track Finishes

PHOTOELECTRIC FINISHES
AT THE RACE TRACK

LRML
Source

RCPT360
Control

You can always bet on a dependable performance from Autotron Photoelectric Controls. The finish of the race illustrated above appears to be a three-way tie for first place, but if one of the cars has even the slightest edge, the Photoelectric Control will make it a matter of record. A Timing Control converts a very short light beam interruption into a slightly longer pulse to give positive relay closure for a "photo finish". Photoelectric Controls are the most dependable and accurate equipment available to separate the winner from the others. Autotron Photoelectric Controls spot the winner equally well day or night, rain or shine, whatever conditions prevail.

Autotron

Appendix A
Conversions and Constants

Length

    1 m = 39.37 in.   1 in. = 0.0833 ft
         = 3.281 ft   = 25.4 mm
         = 1000 mm   = 2.54 cm
         = 100 cm   = 1000 mils
         = 1.057×10^{-16} light-years

Time

    1 year = 365.242 days
            = 8765.813 hours
            = 31,556,926 seconds

Angles

    1 degree = 0.175 rad
               = 1.0 in. at 5 ft
               = 1.0 ft at 60 ft

Speed

    100 ft/min = 0.5080 m/s
                = 1.136 mi/h
                = 20 in./s
                = 0.020 in./ms

    Light = 186,292.397 mi/s
          = 299,792,458 m/s

[Speed]

> 1 in/s = 5.000 ft/min
> = 0.0568 mi/h
> = 25.4 mm/s
> = 152.7 furlongs/fortnight

> Sound in air = 344 m/s (at 20°C)
> = 1129 ft/sec

Energy

> 1 kWh = 3413 Btu
> = 860.2 kcal
> = 3600 kJ

Power

> 1 hp = 2545 Btu/s
> = 550 ft-lb/s
> = 746 W

Constants

> e 2.7182818284
> π = 3.1415926535
> $\pi!$ = 7.1880827258

Basic Equations

> $V = IR$
> $W = VI$
> $i = C \, (dv/dt)$
> $v = L \, (di/dt)$

Light

> 10,000 fc = 107,639 lux
> = 10.7639 lm/cm^2
> = 0.160 W/cm^2 (650 nm)
> = 0.016 W/cm^2 (555 nm)

> 1 W/m^2 = 673 lux (555 nm)
> = 67 lux (650 nm)
> = 17 lux (2860 K)

Solar radiation at earth's mean distance = 1.35 kW/m^2 (6000 K)

Appendix B
Test Standards and References

The following tests and standards are available from the organizations listed below.

ANSI   Publications Service
American National Standards Institute
1430 Broadway
New York, NY 10018
(212) 354-3300

CSA   Sales Group
Canadian Standards Association
178 Rexdale Blvd.
Rexdale, Ontario, Canada M9W 1R3
(416) 747-4044

DOD   Superintendent of Documents
U.S. Government Printing Office
Washington, DC 20402
(202) 783-3238

EIA   Standards Sales Office
Electronic Industries Association
2001 Eye Street, N.W.
Washington, DC 20006

IEC   Sales Department of Central Office—IEC
International Electrotechnical Commission
3 Ru De Varembe
1211 Geneva 20
Switzerland
TELEX: 845-28-872

IEEE   Service Center, Publications Sales Dept.
          Institute of Electrical and Electronics Engineers, Inc.
          445 Hoes Lane
          Piscataway, NJ 08854
          (201) 981-0060

NEMA   Standards Publications Editor
          National Electrical Manufacturers Association
          2101 L Street, N.W.
          Washington, DC 20037

OPCON   Product Marketing Manager
          Opcon, Inc.
          720 80th St. S.W.
          Everett, WA 98203-6299
          (206) 353-0900

UL   Publications Stock
          Underwriters Laboratories, Inc.
          333 Pfingston Road
          Northbrook, IL 60062

Sunlight Immunity

1. *IES Lighting Handbook*, 5th ed. Illumination Engineering Society, New York, 1978, pp. 7-5, 8-73.
2. Qualification Test Standard (QTS)-106.1, *Sunlight Immunity*, Opcon, Inc.

Fluorescent Light Immunity

1. *IES Lighting Handbook*, 5th ed. Illumination Engineering Society, New York, 1978, p. 8-27.
2. QTS-106.2, *Fluorescent Light Immunity*, Opcon, Inc.

Excess Gain Specification

1. QTS-106.4, *Excess Gain Specification*, Opcon, Inc.

Field of View and Sensing Zone Specification

1. QTS-106.5, *Field of View and Sensing Zone Specification*, Opcon, Inc.

Showering Arc Immunity

1.   ICS 1-109, *Tests and Test Procedures*, NEMA.
2.   QTS-104.2, *Chattering Relay Immunity*, Opcon, Inc.

Supply Ripple Voltage Immunity

1.   MIL-STD-462 Method CS01, *Conducted Susceptibility, 30 Hz to 50 kHz, Power Leads*, DOD.
2.   MIL-STD-462 Method CS02, *Conducted Susceptibility, 50 kHz to 400 MHz, Power Leads*, DOD.
3.   QTS-104.1, *DC Supply Ripple Voltage*, Opcon, Inc.

Radio-Frequency Noise Immunity

1.   MIL-STD-462 Method RS03, *Radiated Susceptibility, 14 kHz to 10 GHz, Electric Field*, DOD.
2.   MIL-STD-462 Method RS04, *Radiated Susceptibility, 14 kHz to 30 MHz, Electric Field*, DOD.
3.   QTS-104.5, *5 Watt Communications Transceivers*, Opcon, Inc.

Transient Voltage Spike Immunity

1.   MIL-STD-462 Method CS06, *Conducted Susceptibility, Spike, Power Leads*, DOD.
2.   C62.41-1980, *IEEE Guide for Surge Voltages in Low-Voltage AC Power Circuits*, ANSI/IEEE.
3.   ICS 1-109.22, *Tests and Test Procedures*, NEMA.

Output Rating

1.   RS-443, *Standard for Solid State Relays*, EIA/NARM.
2.   C37.90-78, *Relays and Relay Systems Associated with Electrical Apparatus*, ANSI.
3.   ICS 2-125, *Contacts for Control Circuit Devices*, NEMA.
4.   QTS-103.6, *Solid State Outputs*, Opcon, Inc.

Enclosures

1.   Pub. No. 250-1979, *Enclosures for Electrical Equipment (1000 Volts or Less)*, NEMA.
2.   QTS-101.2, *High Temperature and Pressure Wash Down*, Opcon, Inc.

Temperature and Humidity

1. MIL-STD-810D Method 502.1, *High Temperature*, DOD.
2. MIL-STD-810D Method 502.2, *Low Temperature*, DOD.
3. QTS-103.1, *Temperature*, Opcon, Inc.
4. MIL-STD-202E Method 103B, *Humidity (Steady State)*, DOD.
5. MIL-STD-810D Method 507.2, *Humidity*, DOD.
6. QTS-103.2, *Humidity*, Opcon, Inc.

Vibration

1. MIL-STD-202E Method 204C, *Vibration, High Frequency*, DOD.
2. MIL-STD-810D Method 514.3, *Vibration*, DOD.
3. QTS-102.2, *Vibration*, Opcon, Inc.

Shock

1. MIL-STD-202E Method 213B, *Shock (Specified Pulse)*, DOD.
2. MIL-STD-202E Method 203B, *Random Drop*, DOD.
3. MIL-STD-810D Method 516.3, *Shock*, DOD.
4. QTS-102.1, *Shock*, Opcon, Inc.

Response-Time Measurement

1. ICS-2-229.45, *Rate of Operating Verification for Proximity Switches*, NEMA.
2. QTS-106.3, *Response Time Measurement*, Opcon, Inc.

Safety Standards

1. UL-508, *Industrial Control Equipment*, UL.
2. C22.2 No. 0-M1982, *General Requirements—Canadian Electrical Code Part II*, CSA.
3. C22.2 No. 142-M1983, *Process Control Equipment*, CSA.
4. *National Electrical Code*.
5. R15.06-1986, *American National Standard for Industrial Robots and Robot Systems—Safety Requirements*, ANSI/RIA.

Appendix C

Resistance to Chemical Attack

Material	Trade name	Attacked by	Resistant to
ABS	Cycolac Lustran	Conc. sulfuric acid Conc. nitric acid Esters Ketones Ethylene dichloride	Aqueous acids Aqueous alkalis Conc. phosphoric acid Conc. hydrochloric acid Alcohols Animal oils Mineral oils Vegetable oils
Acrylic	Plexiglas Lucite	Paint thinner Turpentine Alcohol Acetone Toluene Esters Ketones Aromatic hydro-carbons Chlorinated hydro-carbon Concentrated acids Carbon tetrachlor-ide	Household cleaners VM&P naphtha Kerosene Inortanic alkalis Weak acids Hexane Octane Aliphatic hydrocar-bons

Material	Trade name	Attacked by	Resistant to
Nylon	Zytel	Phenols Formic acid Strong mineral acids Strong oxidizing agents	Esters Ketones Alcohols Hydrocarbons Alkalis to pH 11 Ethers Freons
Polycarbonate	Lexan Merlon	Strong acids Strong alkalis Organic solvents Fuels	Weak acids Weak alkalis
Polyester	Valox Celanex	Strong acids Strong bases Ethylene dichloride Low-molecular-weight ketones Aromatic hydrocarbons	Gasoline Alcohols Glycols High-molecular-weight ketones Dilute acids and bases Aliphatic hydrocarbons Oils and fats Carbon tetrachloride
P.P.O.	Noryl	Heptane Benzene Toluene MEK Methelene chloride Aromatic hydrocarbons Halogenated hydrocarbons	Strong acids Strong bases Oils Detergents Clorox

Cycolac is a trademark of Borg Warner Chemicals.
Lustran is a trademark of Monsanto Plastics and Resins Co.
Plexiglas is a trademark of Rohm and Haas Co.
Lucite and Zytel are trademarks of E. I. du Pont de Nemours & Co.
Lexan, Noryl, and Valox are trademarks of General Electric Co.
Celanex is a trademark of Celanese Plastics.
Clorox is a trademark of The Clorox Company.

References

Bradbury, Ray (1953) "Fahrenheit 451," Simon and Schuster, New York.

Dolan, T. J., and Murray, W. M. (1950). Photoelasticity, in M. Hetenyi, *Handbook of Experimental Stress Analysis*. John Wiley & Sons, Inc., New York, pp. 828–860.

Halliday, D., and Resnick, R. (1978). *Physics*. John Wiley & Sons, Inc., New York.

Rock, Irvin (1984). Lightness Constancy, *Perception*. W. H. Freeman and Company, Publishers, New York, pp. 32–36.

Sieppel, R. G. (1981). Optical Fibers, Cables, and Coupling, *Optoelectronics*. Reston Publishing Co., Inc., Reston, Va., pp. 143–148.

Shortley, G., and Williams, D. (1971). Quantum Properties of Radiation, *Elements of Physics*. Prentice-Hall, Inc., Englewood Cliffs, N.J., pp. 843–846.

Index